KENTUCKY
TOTAL ECLIPSE GUIDE

Official Commemorative 2024 Keepsake Guidebook

2024 Total Eclipse State Guide Series

Aaron Linsdau

SASTRUGI PRESS

JACKSON HOLE

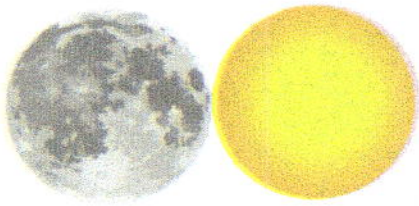

Sastrugi Press / Published by arrangement with the author

Kentucky Total Eclipse Guide: Official Commemorative 2024 Keepsake Guidebook

Sastrugi Press
PO Box 1297, Jackson, WY 83001, United States
www.sastrugipress.com
Quantity sales: Special discounts are available on quantity purchases by corporations, associations, and others. For details, contact the publisher at the address above.

Library of Congress Catalog-in-Publication Data
Library of Congress Control Number: 2019949787
Linsdau, Aaron
Kentucky Total Eclipse Guide / Aaron Linsdau-1st United States edition
p. cm.
1. Nature 2. Astronomy 3. Travel 4. Photography
Summary: Learn everything you need to know about viewing, experiencing, and photographing the total eclipse in Kentucky on April 8, 2024.

ISBN-13: 978-1-64922-318-0 (paperback)

508.4—dc23

All photography, maps and artwork by the author, except as noted.
00057
10 9 8 7 6 5 4 3

Contents

Introduction

Thank you for purchasing this book. It has everything you need to know about the total eclipse in Kentucky on April 8, 2024.

A total eclipse passing through the United States is a rare event. The last US total eclipse was in 2017. It traveled from Oregon to South Carolina. The last American total eclipse prior to that was in 1979!

The next total eclipse over the US will not be until April 8, 2024. It will pass over Texas, the Midwest, and on to Maine. After that, the next coast-to-coast total eclipse will be in 2045.

It's imperative to make travel plans early. You will be amazed at the number of people swarming to the total eclipse path. Some might say watching a partial versus a total eclipse is a similar experience. It's not.

This book is written for Kentucky visitors and anyone else viewing the eclipse. You will find general planning, viewing, and photography information inside. Should you travel to the eclipse path in Kentucky in April, be prepared for an epic trip. The estimates based on the 2017 eclipse suggest that millions will converge on Kentucky.

Some hotels in the communities and cities along the path of totality in Kentucky have already been contacted by people to make reservations. Finding lodging along the eclipse path may be a major challenge.

Resources will be stretched far beyond the normal limits. Think gas lines from the late 1970s. It may be likely that traffic along highways will come to a complete standstill during this event. Be prepared with backup supplies.

Many smaller Kentucky towns are far from any major city. Kentucky country roads can be slow. Please obey posted speed limits for the safety of everyone. Be cautious about believing a map application's estimate of travel time in Kentucky.

People in communities along the path of the total eclipse may rent out properties for this event. With this major celestial spectacle in the spring of 2024, be assured that Kentucky "hasn't seen anything yet."

Is this to say to avoid Kentucky or other areas during the eclipse? Not at all! This guidebook provides ideas for interesting, alternative, and memorable locations to see the eclipse. It will be too late to rush to a better spot once the eclipse begins. Law enforcement will be out to help drivers reconsider speeding.

Please be patient and careful. There will be a large rush of people from all over the world, converging on Kentucky to enjoy the total eclipse. Be mindful of other drivers on eclipse weekend, as they may not be familiar with Kentucky roads.

You should feel compelled to play hooky on April 8. Ask for the day off. Take your kids out of school. They'll be adults before the next chance to see a total eclipse over America. Create family memories that will last a lifetime. Sastrugi Press does not normally advocate skipping school or work. Make an exception because this is too big an event to miss.

Wherever you plan to be along the total eclipse path, leave early and remember your eclipse glasses. People from all around the planet will converge on Kentucky. Be good to your fellow humans and be safe. We all want to enjoy this spectacular show.

Visit www.sastrugipress.com/eclipse for the latest updates for this state eclipse book series.

Author Information

Polar explorer and motivational speaker Aaron Linsdau's first book, *Antarctic Tears*, is an emotional journey into the heart of Antarctica. He ate two sticks of butter every day to survive. Aaron coughed up blood early in the expedition and struggled with equipment failures. Despite the endless difficulties, he set a world record for surviving the longest solo expedition to the South Pole.

Aaron teaches how to build resilience to overcome adversity by managing attitude. He shares his techniques for overcoming adrenaline burnout and constant overload. He inspires audiences to face their challenges with a new perspective. As a motivational speaker, Aaron talks about courage, resilience, attitude, safety, and risk. He hopes that you will be inspired and have an enjoyable time watching the total eclipse in Kentucky.

Visit his websites at www.aaronlinsdau.com or www.ncexped.com.

All About Kentucky

Overview of Kentucky

Kentucky has something to offer just about anyone. Whether it's history that interests you or sports, scenery, or bourbon, Kentucky is a great place to visit and will most likely make you want to come back for more.

Native American cultures have been calling Kentucky home since at least 1800 BCE, leaving their mark on the history of all areas of the Commonwealth. Before that, mammoths and ancient fauna were drawn to the salt deposits that still dot the state, which can be visited and explored today at the Big Bone Lick State Park in Union. Later, white settlers traveled through the Cumberland Gap from Virginia, led by intrepid explorers like Daniel Boone. Their trail can still be

seen at the Cumberland Gap National Historical Park in Middlesboro. Starting as a county of Virginia, Kentucky joined the United States in 1792 as the fifteenth state.

Kentucky was the birthplace of both United States President Abraham Lincoln (Hodgenville) and Confederate States President Jefferson Davis (Fairview), and while technically Kentucky voted to join the Confederacy, the pro-Union feelings of most of the population kept it neutral and Kentuckians fought on both sides of the Civil War. Minor battles of the war were fought at Richmond, Perryville, and Middle Creek, which are open to visitors.

The first thing most people think of with Kentucky is horses. Held at Churchill Downs racetrack every May, the Kentucky Derby is an epic event that draws enthusiasts from around the world. Many people host viewing parties at homes, bars, or restaurants around Kentucky so you can still watch and enjoy the race while wearing your fanciest clothes and sipping a mint julep even if you can't get to Louisville. In addition, horse farm tours in the horse country between Louisville and Lexington are available. The Kentucky Horse Center in Lexington gives an outstanding history of horses and the horse industry as well as shows and demonstrations and a chance to meet famous retired horses.

Of course, no trip to Louisville is complete without a visit to where Louisville Slugger bats were made. While most of the bat production is no longer carried out at the original factory, you can still explore the history of baseball and the bats that make it possible. Visitors can watch as bats are made on the premises.

Both Kentucky's flagship universities, University of Louisville and University of Kentucky in Lexington, have storied traditions of basketball greatness. If you can get tickets (most of them are season passes, so it can be difficult), try catching a game . College football is almost as popular and somewhat easier to acquire tickets, so if you happen to visit in the fall, check the schedule for both universities.

Motorsports fans will also enjoy Kentucky. NASCAR, IRL, and other racing leagues have major events at Kentucky Speedway in Sparta. Corvettes are built at the Chevrolet plant in Bowling Green, and the Corvette Museum should not be missed by anyone with an interest in cars.

The other industry most people remember about Kentucky is bourbon. Bourbon has been distilled in Kentucky since the eighteenth century and is still a vital part of the economy. Distilleries are located throughout the state but are mainly in the central area. There are dozens of distilleries, including famous names like Maker's Mark and Buffalo Trace, which offer tours and tastings.

For all of its attractions, Kentucky's greatest treasure is scenery. From the mountains and forests of the Appalachian chain in eastern Kentucky to the gently rolling hills of the Bluegrass region, to the prairies, lakes, and caves out west, Kentucky has something for everyone. Cumberland Falls, near Corbin, is famous for its moonbows, and the huge lakes at Kentucky Lake State Park have to be seen to be believed. You might see everything from black bears to bald eagles, elk, and more. Kentucky's state resort parks provide quality hotels at most of its state parks, which are reasonably priced and have great amenities.

Hotels and Motels During the Eclipse

Once excitement of the total eclipse over Kentucky spreads, rooms will become scarce. Many hotels in towns along the path of totality in western states sold out for a year or more during the 2017 total eclipse. Kentucky is not alone in this challenge. Hotels all along the path of totality will sell out in anticipation of the 2024 total eclipse.

What does this mean for eclipse visitors? Lodging and room rentals in eclipse towns will be at a massive premium. Does that mean all hope is lost to find a place to stay? Not at all. But you will have to be creative. There will be few, if any, hotel rooms available in these eclipse cities by early 2024. Accommodations in the cities and towns along the path of the eclipse will be difficult to come by.

In summer 2017, the author searched on Hotels.com for rooms along the 2017 total eclipse path on the weekend of August 21 and found many major cities sold out. Once word of the 2024 eclipse spreads, room rates will increase and availability will drop.

Search for rooms farther away from the eclipse path. If you are willing to stay in cities outside the eclipse path, you will have better success at finding rooms. As the eclipse approaches, people will book rooms farther from the totality path. By early spring, rooms in cities near the total eclipse path may be unavailable. The effect of this event will be felt across Kentucky and the rest of the United States.

Think regionally when looking for rooms. Be prepared to search far and wide during this major event. If a five-hour drive is manageable, your lodging options greatly expand, but it also increases your travel risk.

Internet Rentals

To find rooms to stay in towns along the eclipse path, try a web service such as Airbnb.com. Note that some people rent out rooms or homes illegally, against zoning regulations. Cities will feel the crunch of inquiries early due to others who experienced the 2017 eclipse.

If cities fully enforce zoning laws, authorities may prevent your weekend home rental. Online home rentals during the eclipse will be a target for rental scams. People from out of the area steal photos and descriptions, then post the home for rent. You send your check or wire money to a "rental agent" then show up to find you have been scammed. If the deal sounds strange or too good to be true, run away.

Camping

If you can book a campsite, do it as soon as you can. Do not wait. All areas in the national forests are first-come, first-served. Forest roads may be packed. Expect all areas to be swarming with people. Show up early to stake out your spot. Consider staying farther away and driving early on April 8.

Please respect private land too. Kentucky folks don't take kindly to people overrunning their property without permission. In a big state with millions of residents, people are very protective, but they're friendly, too. You never know what you might be able to arrange with a smile and a bit of money.

This all said, there are plenty of camping opportunities throughout Kentucky. You don't have to sleep exactly on the eclipse path. If you're ready to rough it, there are national forest camping options.

Government agencies will meet years in advance to talk about how to manage the influx of people. Every possible government agency will be working full time to enforce the various rules and regulations.

National Parks and Monuments

Finding a camping site at any state park, national park, or national monument along the eclipse path in Kentucky will be challenging. To watch the eclipse from any location, you do not have to sleep in it. You just need to drive there in the morning.

Law enforcement will be present on the eclipse weekend. Hundreds of thousands of people are expected in the region. Parking may overflow. It will make parking lots and lines on Black Friday at the mall look uncrowded. For an event of this magnitude, find your location as early as possible.

The first sentence of the national parks mission statement is:

"The National Park Service preserves unimpaired the natural and cultural resources and values of the national park system for the enjoyment, education, and inspiration of this and future generations."

Roadside camping (sleeping in your car) is not allowed in national monuments or parks. Park facilities are only designed to handle so many people per day. Water, trash collection, and toilets can only withstand so much. If you notice trash on the ground, take a moment to throw it away. Protect your national park and help out. Rangers are diligent and hardworking but they can only do so much to manage the expected crowds.

National and State Forests

There are national and state forest options in Kentucky. They all have camping opportunities. The forest service manages undeveloped and primitive campsites. Be sure to check for any fire restrictions. Check with individual agencies for last-minute information and regulations. The forest service requires proper food storage. Plan to purchase food and water before choosing your campsite. Below is a partial list of national forests in Kentucky:

Daniel Boone National Forest: https://www.fs.usda.gov/dbnf/
George Washington and Jefferson National Forests
Knox National Forest
Clifty Wilderness

Backcountry service roads abound in Kentucky. Maps for forests are available at local visitor centers and bookstores. This book's website has digital copies of some forest maps.

Printed national forest maps are large and detailed. They have illustrated road paths, connections, and other vital travel information not available on digital device maps. Viewing digital maps on your smartphone or mobile pad is difficult. If you plan to camp in the forest, a real paper map is a wise investment.

Camping in federal wilderness areas is also allowed. Those areas afford the ultimate backcountry experience. However, be aware that no vehicle travel is allowed in the specially designated areas. This ban includes: vehicles, bikes, hang gliders, and drones. You can travel only on foot or with pack animals.

Sleep in Your Car

Countless RVs, campers, trucks, cars, and motorcycles will flood Kentucky. Sleeping in your car with friends is tolerable. Doing so with unadventurous spouses or children is another matter.

Do not be caught along the path of the total eclipse without some sort of plan, especially in the bigger cities of Kentucky. The whole path of totality will fill with people on April 8.

Useful Local Webcams

Local webcams are handy to make last-minute travel decisions. Modern webcams are sensitive enough to show headlights at night. Use them to determine if there are issues before traveling out. Eclipse traffic will add to the morning commuter traffic.

There are smartphone applications which are useful to check webcams in many locations. Consult your device's app store for the latest updates. Whether you use an app or computer, an Internet search will reveal many handy webcams for your eclipse planning.

Weather

It's all about the weather during the eclipse. Nothing else will matter if the sky is cloudy. You can be nearly anywhere along the path in Kentucky and catch a view of the event when traffic comes to a standstill. But if there's a cloud cover forecast, seriously reconsider your viewing location.

Travel early wherever you plan to go. Attempting to change locations an hour before the eclipse due to weather will likely cause you to miss the event. Kentucky country roads can be narrow and slow. The number of vehicles will cause unexpected backups.

Modern Forecasts

Use a smartphone application to check the up-to-date weather. Wunderground is a good application and has relatively reliable forecasts for the region. The hourly forecast for the same day has been rather accurate for the last two years. The below discussion refers to features found in the Wunderground app. However, any application with detailed weather views will improve your eclipse forecasting skills.

Cloud Cover Forecast

The most useful forecast view is the visible and infrared cloud-coverage map. Avoid downloading this app the night before and trying to learn how to read it. Practice reading them at home. It's imperative to understand how to interpret the maps early.

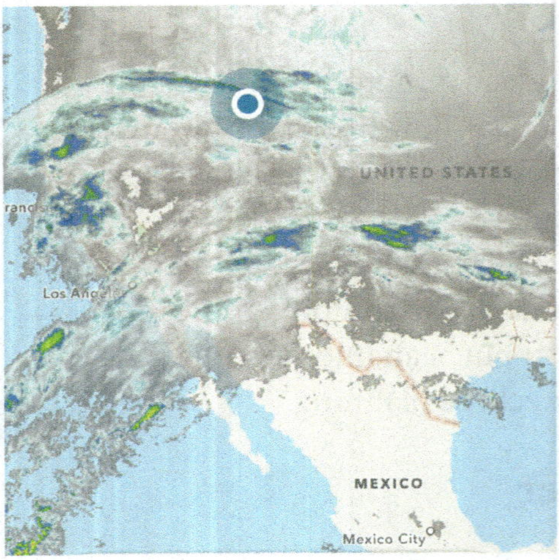

Infrared cloud map showing the worst case eclipse cloud cover. Courtesy of National Weather Service.

All cloud cover, night or day, will appear on an infrared map. Warm, low-altitude clouds are shown in white and gray. High-altitude cold clouds are displayed in shades of green, yellow, red, and purple. Anything other than a clear map spells eclipse-viewing problems.

To improve your weather guess, use the animated viewer of the cloud cover. It will give you a sense of cloud motion. You can discern whether clouds or rain are moving toward, away from, or circulating around your location.

Normal Kentucky Weather Pattern

Due to the direction of the jet stream, most weather travels across the Pacific Ocean, through the western states, over the Rockies, and then into Kentucky. On occasion, weather can approach from Canada or Mexico. Due to the nature of the storms from the Arctic, weather in Kentucky can be unpredictable.

The common weather pattern in April is slightly warm in the afternoon and mildly cool in the evenings. Passing cold fronts in spring can bring unexpected cloud cover and rains.

Historically, Kentucky tends to have moderate cloud cover during April. Prepare to make adjustments. If anything other than clear skies are predicted, drive to other parts of Indiana, Missouri, or Illinois.

Be aware of tornadoes in Kentucky. Although the peak tornado season is June, there have been many recorded tornadoes in April. Pay attention to the weather forecast. If dangerous weather is predicted, your main concern should be safety rather than chasing an eclipse.

Consider that slow-moving clouds can obscure the sun for far longer than the four-minute duration of the totality. The time of totality is so short that you do not want to risk it. Missing it due to a single cloud will be a major disappointment.

Local Eclipse Weather Forecasts

Local town and city newspapers, radio, and television stations around Kentucky will have a weekend edition with articles discussing the eclipse weather. However, conditions change unpredictably in Kentucky. A three-day forecast for April may be incorrect.

Finding the Right Location to View Eclipse Effects

One of the peculiarities of total eclipses is that the entire show is not only in the sky. There are other unusual effects seen during the total eclipse that are worth looking for.

The first effect to watch for is the crescent moon shapes created from leaf shadows on the ground. They're best viewed on a sidewalk or asphalt. They can only be observed during the partial eclipse. The other effect that is worth watching for is the shadow bands or "snakes" as they're commonly called.

Shadow banding is seen right before and after the totality takes place. They're best observed on smooth, plain-colored surfaces. If you plan to be in the forest for the eclipse, you may struggle to see the bands but will likely see crescent shadows all around on the ground.

One of the supreme challenges with all of these effects is choosing what to watch. You can see the crescent shadows in the hour before and after the totality but shadow banding happens before or after totality. It is more difficult to look away from the eclipse than you think.

Road Closures and Traffic

Highways connecting various Kentucky towns in the total eclipse path will be heavily impacted on the weekend before and day of the eclipse. As was found in the 2017 total eclipse, there is no way to predict which areas will be impacted.

Planning ahead is essential to give you the best opportunity to enjoy the eclipse without the nightmare of being stuck in traffic for hours on end. The traffic in states during the 2017 eclipse was stunning, so imagine what it will be like for Kentucky in 2024.

Update yourself with the latest road report information from the Kentucky road condition website:

https://transportation.ky.gov/sites/GoKY/home

It's imperative to plan early and have one if not more backup plans in case of difficult travel conditions. April weather is unpredictable and variable.

If you believe it's necessary to leave a town to watch the eclipse, do so the night before or extremely early in the morning. RVs are common, and trains of them crawl through popular areas.

Communication Information

Cellular Phones

Cellular "cell" phone service in some remote Kentucky locations may be problematic. Most of the time there is good coverage along the main highways and interstates. However, even along major thoroughfares, there can be little or no coverage.

It's possible to find zones where text messages will send when phone calls are impossible. If you cannot make a phone call, the chance of having data coverage for web surfing or e-mail is low.

Please look up any information or communicate what you need before departing from the main roads around Kentucky. Bureau of Land Management (BLM) areas sometimes have coverage. Planned to be self-contained. Plan for your cell phone not to connect.

You may find yourself out of cell service. With a large number of cell users in a concentrated area, coverage and data speed may collapse as well. Search on the phrase "cell phone coverage breathing".

Wilderness and Forest Safety

All Kentucky forest and wilderness areas are full of wild animals. Although beautiful, wild animals can be dangerous. They can easily injure or kill people, as they are far more powerful than humans. Do not try to feed any wild animals, including squirrels, foxes, and chipmunks, as they can carry diseases. These suggestions apply to all public lands.

Ticks

Ticks exist all across the United States, but not all species transmit disease. Ticks cannot fly or jump, but they climb grasses in shrubs in order to attach to people or animals that pass by. Ticks feed on the blood of their host. In doing so, they can transmit potentially life-threatening diseases such as Lyme disease.

Spiders

Although the chance of encountering a venomous spider is low, it is not uncommon to encounter them. The two dangerous spiders in the state are the black widow and brown recluse. Should you encounter either species, simply leave it alone. If you are bitten, seek immediate medical attention, as their toxin can be life-threatening.

Venomous Snakes

There are multiple species of venomous snakes in Kentucky including the Copperhead, Western Cottonmouth, Western Pygmy Rattlesnake, and Timber Rattlesnake. Although these reptiles are not generally aggressive, they can strike when provoked or threatened. Of the approximately 8,000 people annually bitten by venomous snakes in the United States, ten to fifteen people die according to the U.S. Food and Drug Administration.

The best way to avoid rattlesnake encounters is to be mindful of your environment. Do not place your hands or feet in locations where you cannot clearly see the surroundings. Avoid heavy brush or tall weeds where snakes hide during the day. Step on a log or rock rather than over it, as a hidden snake might be on the other side. Rattlesnakes may not make any noise before striking.

Avoid handling all snakes. Should you be bitten, stay calm and call 911 or emergency dispatch as soon as possible. Transport the victim to the nearest medical facility immediately. Rapid professional treatment is the best way to manage rattlesnake bites. Refer to US Forest Service and professional medical texts for more information on managing rattlesnakes injuries.

Bears

The forests of Kentucky are potentially home to black bears. Though they are listed rarely seen, they have been sighted in the state. Although they often appear docile, they can become aggressive if threatened. In the unlikely event of an attack, fight back against the bear. Use whatever you have at your disposal to defend yourself. Report all negative or aggressive bears to the local authorities.

If a bear hears you, it will usually vacate the area. Bear charges are often caused by unexpected and surprise encounters. Noise is the best defense to avoid surprising bears. Regularly clap, make noise, and talk loudly. The Kentucky Department of Fish and Wildlife Resources website has more specific information on safety and food management in bear country at https://fw.ky.gov/Wildlife/Pages/Black-Bears.aspx.

It is recommended to stay one hundred yards (300 feet) away from all bears. They are exciting to see but need their space. Refer to current forest or park regulations for more safety information.

Mountain Lions

Though declared to be extinct in the state, there have been news reports of mountain lions by various new agencies. If you encounter a mountain lion, do not run. Keep calm, back away slowly, and maintain eye contact. Do all you can to appear larger. Stand upright, raise your arms, or hoist your jacket. Never bend over or crouch down. If attacked, fight back.

Eclipse Day Safety

1. Hydrate

Spring temperatures are usually mild to warm. The excitement of the event can distract you from managing hydration. Drink plenty of water. Consume more than you would at home.

2. Eye Safety time

Use certified eclipse safety glasses at all times when viewing the partial eclipse. Only remove the glasses when the totality happens. Give your eyes time to rest. They can dry out and become irritated. Bring FDA approved eye drops to keep your eyes moist.

3. Sun exposure

Facing at the sun for three hours can result in sunburns. Wear sunglasses and liberally apply sunscreen to avoid sunburns.

4. Eat well

Keep your energy up. Appetite loss is common when traveling. Maintain your normal eating schedule.

5. Prepare for temperature changes

Temperatures will drop rapidly during the eclipse and also once the sun sets. Bring appropriate clothing.

6. Talk with your doctor

If the humidity or heat bothers you talk with your doctor before traveling. Seek professional medical attention for serious symptoms.

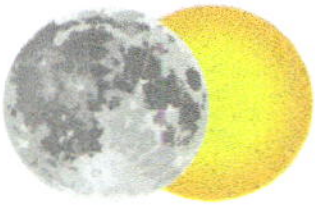

Total eclipse path across the United States (approximate).

Total eclipse path across Kentucky (approximate).

All About Eclipses

How an Eclipse Happens

An eclipse occurs when one celestial body falls in line with another, thus obscuring the sun from view. This occurs much more often than you'd think, considering how many bodies there are in the solar system. For instance, there are over 150 moons in the solar system. On Earth, we have two primary celestial bodies: the sun and the moon. The entire solar system is constantly in motion, with planets orbiting the sun and moons orbiting the planets. These celestial bodies often come into alignment. When these alignments cause the sun to be blocked, it is called an eclipse.

For an eclipse to occur, the sun, Earth, and moon must be in alignment. There are two types of eclipses: solar and lunar. A solar eclipse occurs when the moon obscures the sun. A lunar eclipse occurs when the moon passes through Earth's shadow. Solar eclipses are much more common, as we experience an average of 240 solar eclipses a century compared to an average of 150 lunar eclipses. Despite this, we are more likely to see a lunar eclipse than a solar eclipse. This is due to the visibility of each.

For a solar eclipse to be visible, you have to be in the moon's shadow. The problem with viewing a total eclipse is that the moon casts a small shadow over the world at any given time. You have to be in

a precise location to view a total eclipse. The issue that arises is that most of these locations are inaccessible to most people. Though many would like to see a total solar eclipse, most aren't about to set sail for the middle of the Pacific Ocean. In fact, a solar eclipse is visible in the same place on the world on average every 375 years. This means that if you miss a solar eclipse above your hometown, you're not going to see another one unless you travel or move.

It's much easier to catch a glimpse of a lunar eclipse, even though they occur at a much lower frequency than their solar counterparts. A lunar eclipse darkens the moon for a few hours. This is different than a new moon when it faces away from the sun. During these eclipses, the moon fades and becomes nearly invisible.

Another result of a lunar eclipse is a blood moon. Earth's atmosphere bends a small amount of sunlight onto the moon turning it orange-red. The blood moon is caused by the dawn or dusk light being refracted onto the moon during an eclipse.

Lunar eclipses are much easier to see. Even when the moon is in the shadow of Earth, it's still visible throughout the world because of how much smaller it is than Earth.

Total vs. Partial Eclipse

What is the difference between a partial and total eclipse? A total eclipse of either the sun or the moon will occur only when the sun, Earth, and the moon are aligned in a perfectly straight line. This ensures that either the sun or the moon is partially or completely obscured.

In contrast, a partial eclipse occurs when the alignment of the three celestial bodies is not in a perfectly straight line. These types of eclipses usually result in only a part of either the sun or the moon being obscured. This is often what led to ancient civilizations believing that some form of magical beast or deity was eating the sun or the moon. It appears as though something has taken a bite out of either the sun or the moon during a partial eclipse.

Total eclipses, rarer than partial eclipses, still occur quite often. It's more difficult for people to be in a position to experience such an event firsthand. Total solar eclipses can only be viewed from a small portion of the world that falls into the darkest part of the moon's shadow. Often this happens in the middle of the ocean.

The Moon's Shadow

The moon's shadow is divided into two parts: the umbra and the penumbra. The former is much smaller than the latter, as the umbra is the innermost and darkest part of the shadow. The umbra is thus the central point of the moon's shadow, meaning that it is extremely small in comparison to the entire shadow. For a total solar eclipse to be visible, you need to be directly beneath the umbra of the moon's shadow. This is because that is the only point at which the moon completely blocks the view of the sun.

In contrast, the penumbra is the region of the moon's shadow in which only a portion of the light cast by the sun is obscured. When

Total eclipse shadow 2016 as seen from 1 million miles on the Deep Space Climate Observatory satellite. Courtesy of NASA.

standing in the penumbra, you are viewing the eclipse at an angle. In the penumbra, the moon does not completely block the sun from view. This means that while the event is a total solar eclipse, you'll only see a partial eclipse. The umbra for the April 8 eclipse is over one hundred miles wide. The penumbra will cover much of the United States.

To provide some context, one total solar eclipse we experienced occurred on March 9, 2016, and was visible as a partial eclipse across most of the Pacific Ocean, parts of Asia, and Australia. However, the only place in the world to view this total solar eclipse was in a few parts of Indonesia.

Due to the varied locations and the brief periods for which they're visible, it's difficult to see each and every eclipse that occurs. The umbra of the moon is such a small fraction of the entire shadow and the majority of our planet is comprised of water. Thus, the rarity of being able to view a total solar eclipse increases significantly because it's likely that the umbra will fall over some part of the ocean rather than a populated landmass. There are not total or annular eclipses every month because the moon's orbit is 5.1° off the ecliptic plane of the Earth and sun.

Eclipses Throughout History

Ancient peoples believed eclipses were from the wrath of angry gods, portents of doom and misfortune, or wars between celestial beings. Eclipses have played many roles in cultures, creating myths since the dawn of time. Both solar and lunar eclipses affected societies worldwide. Inspiring fear, curiosity, and the creation of legends, eclipses have cast a long shadow in the collective unconscious of humanity throughout history.

Early Myth & Astronomy

Documented observations of solar eclipses have been found as far back in history as ancient Egyptian and Chinese records. Timekeeping was important to ancient Chinese cultures. Astronomical observations were an integral factor in the Chinese calendar. The first observation of a solar eclipse is found in Chinese records from over 4,000 years ago. Evidence suggests that ancient Egyptian observations may predate those archaic writings.

Many ancient societies, including Roman, Greek and Chinese civilizations, were able to infer and foresee solar eclipses from astronomical data. The sudden and unpredictable nature of solar eclipses had a stressful and intimidating effect on many societies that lacked the scientific insight to accurately predict astronomical events. Relying on the sun for their agricultural livelihood, those societies interpreted solar eclipses as world-threatening disasters.

In ancient Vietnam, solar eclipses were explained as a giant frog eating the sun. The peasantry of ancient Greece believed that an eclipse was the sign of a furious godhead, presenting an omen of wrathful retribution in the form of natural disasters. Other cultures were less speculative in their investigations. The Chinese Song Dynasty scientist Shen Kuo proved the spherical nature of the Earth and heavenly bodies through scientific insight gained by the study of eclipses.

The Eclipse in Native American Mythology

Eclipses have played a significant role in the history of the United States. Before Europeans settled in the Americas, solar eclipses were important astronomical events to Native American cultures. In most native cultures, an eclipse was a particularly bad omen. Both the sun and the moon were regarded as sacred. Viewing an eclipse, or even being outside for the duration of the event, was considered highly taboo by the Navajo culture. During an eclipse, men and women would simply avert their eyes from the sky, acting as though it was not happening.

The Choctaw people had a unique story to explain solar eclipses. Considering the event as the mischievous actions of a black squirrel and its attempt to eat the sun, the Choctaw people would do their best to scare away the cosmic squirrel by making as much noise as possible until the end of the event, at which point cognitive bias would cause them to believe they'd once again averted disaster on an interplanetary scale.

Contemporary American Solar Phenomena

The investigation of solar phenomena in twentieth-century American history had a similarly profound effect on the people of the United States. A total solar eclipse occurring on the sixteenth of June, 1806, engulfed the entire country. It started near modern-day Arizona. It passed across the Midwest, over Ohio, Pennsylvania, New York, Massachusetts, and Connecticut. The 1806 total eclipse was notable for being one of the first publicly advertised solar events. The public was informed beforehand of the astronomical curiosity through a pamphlet written by Andrew Newell entitled *Darkness at Noon, or the Great Solar Eclipse*.

This pamphlet described local circumstances and went into great detail explaining the true nature of the phenomenon, dispelling myth and superstition, and even giving questionable advice on the best methods of viewing the sun during the event. Replete with a short historical record of eclipses through the ages, the *Darkness at Noon* pamphlet is one of the first examples of an attempt to capitalize on the mysterious nature of solar eclipses.

Another notable American solar eclipse occurred on June 8, 1918. Passing over the United States from Washington to Florida, the eclipse was accurately predicted by the U.S. Naval Observatory and heavily documented in the newspapers of the day. Howard Russell Butler, painter and founder of the American Fine Arts Society, painted the eclipse from the U.S. Naval Observatory, immortalizing the event in *The Oregon Eclipse*.

Four more total solar eclipses occurred over the United States in the years 1923, 1925, 1932, and 1954, with another occurring in 1959. The October 2, 1959, solar eclipse began over Boston, Massachusetts. It was a sunrise event that was unviewable from the ground level. Eminent astronomer Jay Pasachoff attributed this event to sparking his interest in the study of astronomy. Studying under Professor Donald Menzel of Williams College, Pasachoff was able to view the event from an airline hired by his professor.

To this day, many myths surround the eclipse. In India, some local customs require fasting. In eastern Africa, eclipses are seen as a danger to pregnant women and young children. Despite the mystery and legend associated with unique and rare astronomical events, eclipses continue to be awe-inspiring. Even in the modern day, eclipses draw out reverential respect for the inexorable passing of celestial bodies. They are a reminder of the intimate relationship between the denizens of Earth and the universe at large.

Present Day Eclipses

The year 2017 brought the world's most-watched total eclipse in history on August 21, when a total solar eclipse crossed the United States. An annular eclipse, a "ring of fire," will pass over the United States in 2023 from Oregon to Texas. Though impressive, it will not

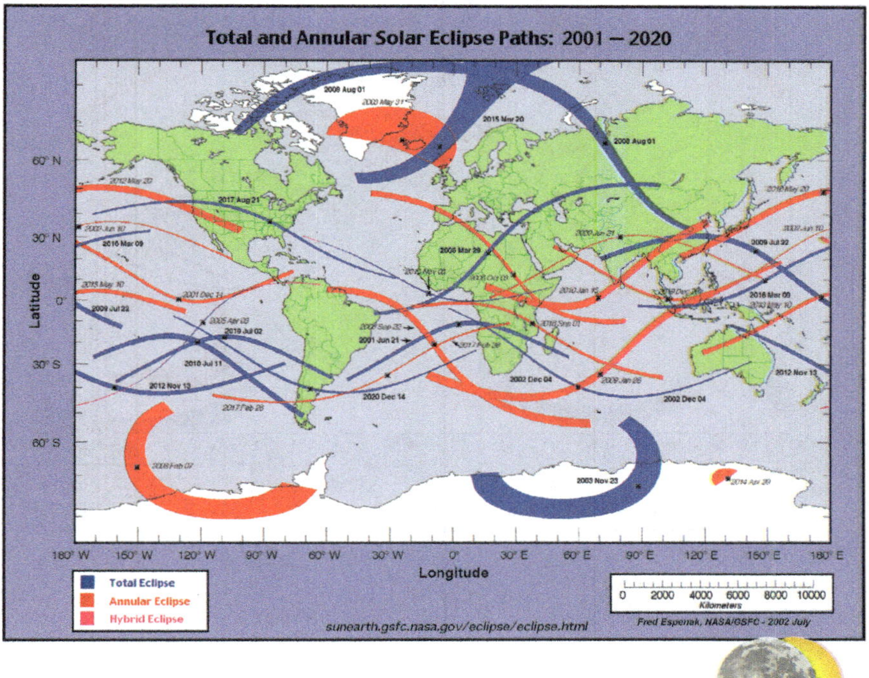

COURTESY OF NASA

compare to the 2024 total eclipse. There is little in nature that equals the spectacle of the sun's corona and seeing stars in the day.

There will be multiple partial, annular, or hybrid eclipses across the world before the 2024 total eclipse. However many are in remote, inaccessible, or potentially dangerous locations on the globe. In 2019 and 2020, Chile and Argentina will experience total eclipses. The next total eclipse after that will occur over Antarctica in 2021. An extremely rare hybrid eclipse will happen in 2023 over the Indian Ocean, Australia, and Indonesia.

The next total solar eclipse viewable from the United States will occur on April 8, 2024. It will be visible in fifteen states: Texas, Oklahoma, Arkansas, Missouri, Tennessee, Kentucky, Illinois, Indiana, Ohio, Pennsylvania, Michigan, New York, Vermont, New Hampshire, and Maine.

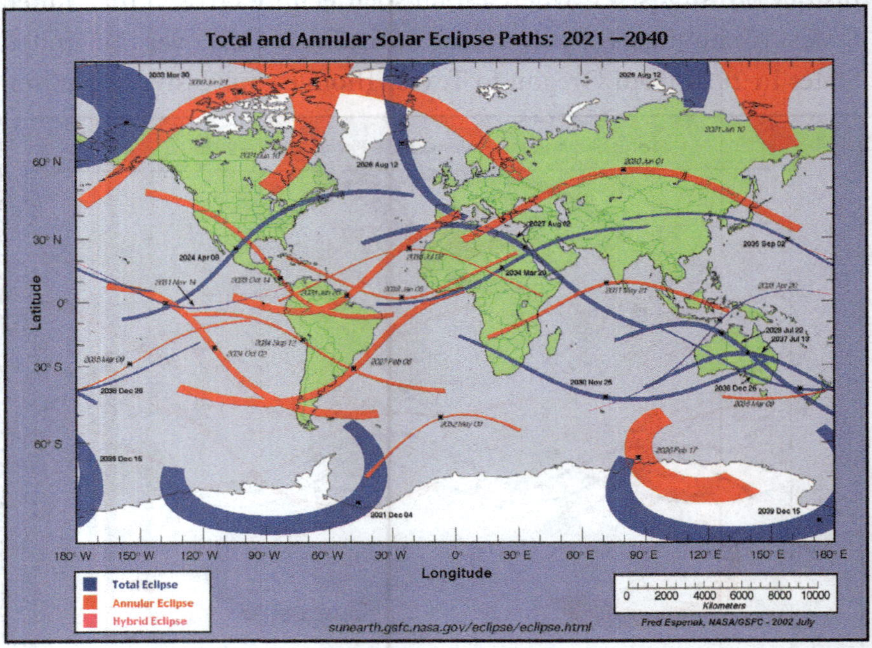

COURTESY OF NASA

Viewing and Photographing the Eclipse

At-home Pinhole Method

Use the pinhole method to view the eclipse safely. It costs little but is the safest technique there is. Take a stiff piece of single-layer cardboard and punch a clean pinhole. Let the sun shine through the pinhole onto another piece of cardboard. That's it!

Never look at the sun through the pinhole. Your back should be toward the sun to protect your eyes. To brighten the image, simply move the back piece of cardboard closer to the pinhole. To see it larger, move the back cardboard farther away. Do not make the pinhole larger. It will only distort the crescent sun.

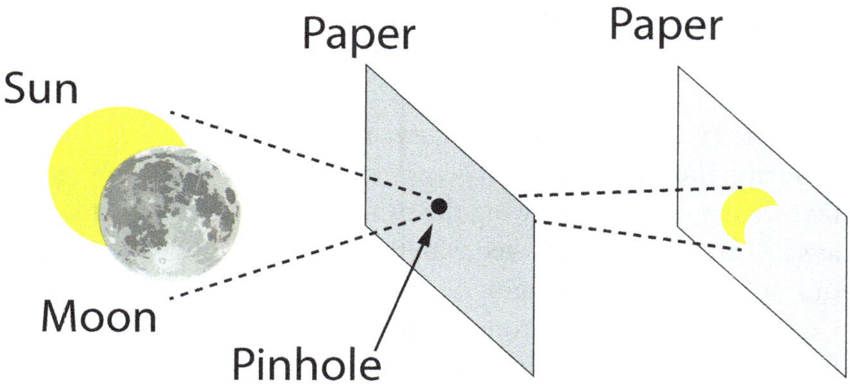

Welding Goggles

Welding goggles that have a rating of fourteen or higher are another useful eclipse viewing tool. The goggles can be used to view the solar eclipse directly. Do not use the goggles to look through binoculars or telescopes, as the goggles could potentially shatter due to intense direct heat. Avoid long periods of gazing with the goggles. Look away every so often. Give your eyes a break.

Solar Filters for Telescopes

The ONLY safe way to view solar eclipses using telescopes or binoculars is to use solar filters. The filters are coated with metal

to diminish the full intensity of the sun. Although the filters can be expensive, it is better to purchase a quality filter rather than an inexpensive one that could shatter or melt from the heat.

The filters attach to the front of the telescope for easy viewing. Remember to give your telescope cooling breaks. Rapid heating can damage your equipment with or without filters attached.

Watch Out for Unsafe Filters

There are several myths surrounding solar filters for eclipse viewing. In order for filters to be safe, they must be specially designed for looking at a solar eclipse. The following are all unsafe for eclipse viewing and can lead to retinal damage: developed colored or chromogenic film, black-and-white negatives such as X-rays, CDs with aluminum, smoked glass, floppy disk covers, black-and-white film with no silver, sunglasses, or polarizing films.

Watch Out for Unsafe Eclipse Glasses

During the 2017 total eclipse, several vendors sold eclipse glasses that were not safe for viewing the sun. Although they were marketed as safe and were even marked with the ISO 12312-2 certification, they did not block eye-damaging visible, infrared, and ultraviolet light. Check the American Astronomical Society's website (eclipse.aas.org) for a list of reputable eclipse glasses vendors.

Viewing with Binoculars

When viewing the eclipse with binoculars, it is important to use solar filters on both lenses until totality. Only then is it safe to remove the filter. As the sun becomes visible after totality, replace the filters for safe viewing. Protect your pupils. Remember to give your binoculars a cool-down break between viewings. They can overheat rapidly from being pointed directly at the sun even with filters attached.

Planning Ahead

There are many things to keep in mind when viewing a total eclipse. It is important to plan ahead to get the most out of this extraordinary experience.

Understanding Sun Position

All compass bearings in this book are true north. All compasses point to Earth's magnetic north. The difference between these two measurements is called magnetic declination. The magnetic declination for Kentucky is:

5° 42' W ± 0° 22' (for Frankfort in 2024)

Adjust the declination from the azimuth bearing as given in the text, and set your compass to that direction.

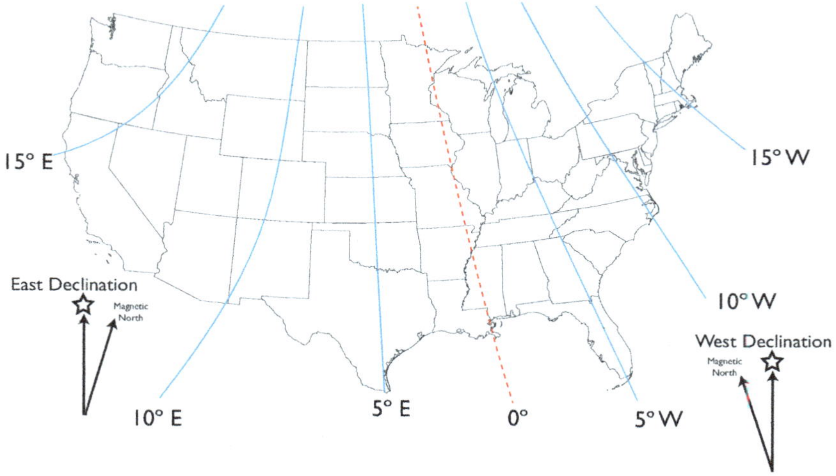

If you purchase a compass with a built-in declination adjustment, you can change the setting once and eliminate the calculations. The Suunto M-3G compass has this correction. A compass with a sighting mirror or wire will help you make a more accurate azimuth sighting.

The Suunto M-3G also has an inclinometer. This allows you to measure the elevation of any object above the horizon. Use this to figure out how high the sun will be above your position.

You can also use a smartphone inclinometer and compass for this purpose. Make sure to calibrate your smartphone's compass before every use, otherwise it might indicate the wrong bearing. Set the smartphone compass for true north to match the book. Understand the compass prior to April 8. There will be little time to guess or search on the Internet. Smartphone and GPS compasses are "sticky." Their compasses don't swing as freely as a magnetic compass does.

The author has used his magnetic compass for azimuth measurements and a smartphone to measure elevation. Combining these two tools will allow you to make the best sightings possible.

Outdoor sporting goods stores in most towns and cities carry compasses. Purchase and practice with a good compass in your hometown well before the event. Take the time to learn how to use it before the day of the eclipse. You do not want to struggle with orienteering basics under pressure.

Sun Azimuth

Azimuth is the compass angle along the horizon, with 0° corresponding to north, and increasing in a clockwise direction. 90° is east, 180° is south, and 270° is west.

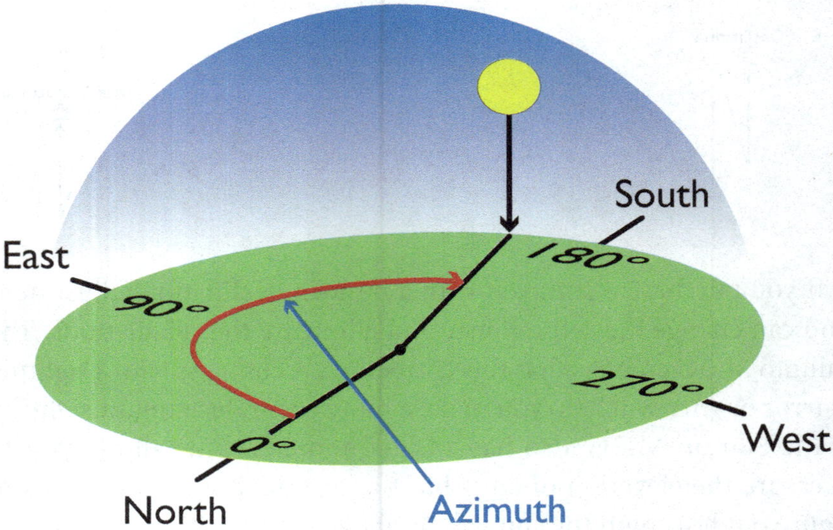

Sun Elevation

Altitude is the sun's angle up from the horizon. A 0° altitude means exactly on the horizon and 90° means "straight up."

Using the sun azimuth and elevation data, you can predict the position of the sun at any given time. Positions given in this book coincide with the time of eclipse totality unless otherwise noted.

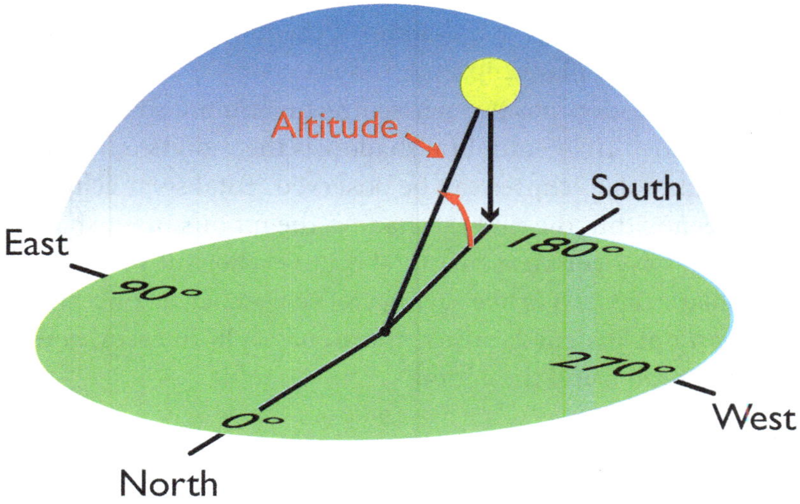

Eclipse Data for Select Kentucky Locations

LOCATION	TOTALITY START (CDT)
BOXVILLE	2:02:14PM
HENDERSON	2:02:39PM
MARION	2:02:20PM
MORGANFIELD	2:01:53PM
PADUCAH	2:00:45PM
STURGIS	2:01:53PM

Eclipse Photography

Photographing an eclipse is an exciting challenge, as the moon's shadow moves near 1,600MPH. There is an element of danger and the pressure of time. Looking at the unfiltered sun through a camera can permanently damage your vision and your camera. If you are unsure, just enjoy the eclipse with specially designed eclipse glasses. Keep a solar filter on your lens during the eclipse and remove for the duration of totality!

Partial Vs. Total Solar Eclipse

To successfully and safely photograph a partial and total eclipse, it is important to understand the difference between the two. A solar eclipse occurs when the moon is positioned between the sun and Earth. The region where the shadow of the moon falls upon Earth's surface is where a solar eclipse is visible.

The moon's shadow has two parts—the penumbral shadow and the umbral shadow. The penumbral shadow is the moon's outer shadow where partial solar eclipses can be observed. Total solar eclipses can only be seen within the umbral shadow, the moon's inner shadow.

You cannot say you've seen a total eclipse when all you saw was a partial solar eclipse. It is like saying you've watched a concert, but in reality, you only listened outside the arena. In both cases, you have missed the drama and the action.

Photographing A Partial And Total Solar Eclipse

Aside from the region where the outer shadow of the moon is cast, a partial solar eclipse is also visible before a total solar eclipse within

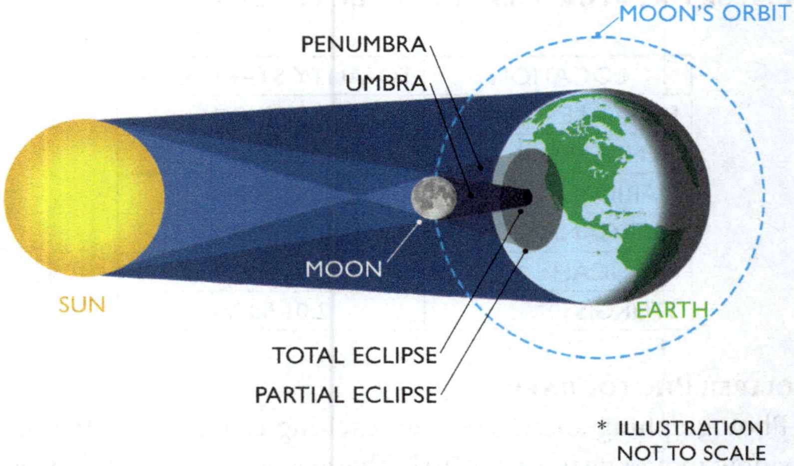

the inner shadow region. In both cases, it is imperative to use a solar filter on the lens for both photography and safety reasons. This is the only difference between taking a partial eclipse and a total eclipse photograph of the sun.

To photograph a total solar eclipse, you must be within the Path of Totality, the surface of the Earth within the moon's umbral shadow.

The Challenge

A total solar eclipse only lasts for a couple of minutes. It is brief, but the scenario it brings is unforgettable. Seeing the radiant sun slowly being covered by darkness gives the spectator a high level of anticipation and indescribable excitement. Once the moon completely covers the sun's radiance, the corona is finally visible. In the darkness, the sun's corona shines, capturing the crowd's full attention. Watching this phenomenon is a breathtaking experience.

Amidst all the noise, cheering, and excitement, you have no more than a few minutes to take a perfect photograph. The key to this is planning. You need to plan, practice, and perfect what you will do when the big moment arrives because there is no replay. The pressure is enormous. You only have a short time to capture the totality and the sun's corona using different exposures.

Plan, Practice, Perfect

It is important to practice photographing before the actual phenomenon arrives. Test your chosen imaging setup for flaws. Rehearse over and over until your body remembers what you will do from the moment you arrive at your chosen spot to the moment you pack up and leave the area.

You will discover potential problems regarding vibrations and focus that you can address immediately. This minimizes the variables that might affect your photographs at the most critical moment.

It's common for experienced eclipse chasers to lose track of what they plan to do. Write down what you expect to do. Practice it time and again. Play annoying, distracting music while you practice. Try photographing in the worst weather possible. Do anything you can to practice under pressure. Eclipse day is not the time to practice.

Once the sun is completely covered, don't just take photographs. Capture the experience and the image of the total solar eclipse in your mind as well. Set up cameras around you to record not just the total solar eclipse but also the excitement and reaction of the crowd.

Eclipse Photography Gear

What do you need to photograph the total eclipse? There are only a few pieces of equipment that you'll need. Preparing to photograph an eclipse successfully takes time. Not only do you have to be skilled and have the right gear, you have to be in the correct place.

Basic Eclipse Photography Equipment

- Solar viewing glasses (verify authenticity)
- Lens solar filter
- Minimum 300mm lens
- Stable tripod that can be tilted to 60° vertical
- High-resolution DSLR
- Spare batteries for everything
- Secondary camera to photograph people, the horizon, etc.
- Remote cable or wireless release

Additional Items

- Video camera
- Video camera tripod
- Quality pair of binoculars
- Solar filters for each binocular lens
- Photo editing software

Equipment to Prepare Before the Big Day

A. Solar viewing glasses

You need a pair of solar viewing glasses as the eclipse approaches.

B. Solar Filter

Partial and total eclipse photography is different from normal photography. Even if only 1% of the sun's surface is visible, it is still approximately 10,000 times brighter than the moon. Before totality, use a solar filter on your lens. Do not look at the sun with your eyes. It can cause irreparable damage to your retinas.

DO NOT leave your camera pointed at the sun without a solar filter attached. The sun will melt the inside of your camera. Think of a magnifying glass used to torch ants and multiply that by one hundred.

C. Lens

To capture the corona's majesty, you need to use a telescope or a telephoto lens. The best focal length, which will give you a large image of the sun's disk, is 400mm and above. You don't want to waste all your efforts by bringing home a small dot where the black disk and majestic corona are supposed to be.

D. Tripod

Bring a stable enough tripod to support your camera properly to avoid unsteady shots and repeated adjustments. Either will ruin your photos. It also needs to be portable in case you need to change locations for a better shot. ***Shut off camera stabilization on a tripod!***

E. Camera

You need to remember to set your camera to its highest resolution to capture all the details. Set your camera to:

- 14-bit RAW is ideal, otherwise
- JPG, Fine compression, Maximum resolution

Bracket your exposures. Shoot at various shutter speeds to capture different brightnesses in the corona. Note that stopping your lens all the way down may not result in the sharpest images.

Choose the lowest possible ISO for the best quality while maintaining a high shutter speed to prevent blurred shots. Set your camera to manual. Do not use AUTO ISO. Your camera will be fooled. The night before, test the focus position of your lens using a bright star or the moon.

Constantly double-check your focus. Be paranoid about this. You can deal with a grainy picture. No amount of Photoshop will fix a blurry, out-of-focus picture.

F. Batteries

Remember to bring fresh batteries! Make sure that you have enough power to capture the most important moments. Swap in fresh batteries thirty minutes before totality.

G. Remote release

Use a wired or wireless remote release to fire the camera's shutter. This will reduce the amount of camera vibration.

H. Video Camera

Run a video camera of yourself. Capture all the things you say and do during the totality. You'll be amazed at your reaction.

I. Photo editing software

You will need quality photo editing software to process your eclipse images. Adobe Lightroom and Photoshop are excellent programs to extract the most out of your images. Become well versed in how to use them at least a month before the eclipse.

J. Smartphone applications

The following smartphone applications will aid in your photography planning: Wunderground, Skyview, Photographer's Ephemeris, Sunrise and Sunset Calculator, SunCalc, and Sun Surveyor among others.

Camera Phones

Smartphone cameras are useful for many things but not eclipse photography. An iPhone 6 camera has a 63° horizontal field of view and is 3264 pixels across. If you attempt to photograph the eclipse, the sun will be a measly 30-40 pixels wide depending on the phone. Digital pinch zoom won't help here. If you want *National Geographic* images, you'll need a serious camera and lens, far beyond any smartphone.

Consider instead using a smartphone to run a time-lapse of the entire event. The sun will be minuscule when shot on a smartphone. Think of something else exciting and interesting do to with it. Purchase a Gorilla Pod, inexpensive tripod, or selfie stick and mount the smartphone somewhere unique.

Also, partial and total eclipse light is strange and ethereal. Consider using that light to take unique pictures of things and people. It's rare and you may have something no one else does.

Focal Length & the Size of Sun

The size of the sun in a photo depends on the lens focal length. A 300mm lens is the recommended minimum on a full-frame (FF) DSLR. Lenses up to this size are relatively inexpensive. For more magnification, use an APS-C (crop) size sensor. Cameras with these sensors provide an advantage by capturing a larger sun.

For the same focal length, an APS-C sensor will provide a greater apparent magnification of any object. As a consequence, a shorter, less expensive lens can be used to capture the same size sun.

The below figure shows the size of the sun on a camera sensor at various focal lengths. As can be seen with the 200mm lens, the sun is quite small. On a full-frame camera at 200mm, the sun will be 371 pixels wide on a Nikon D810, a 36-megapixel body. A lower resolution FF camera will result in an even smaller sun.

Printing a 24-inch image shot on a Nikon D810 with a 200mm lens at a standard 300 pixels per inch results in a small sun. On this size paper, the sun will be a miserly 1.25 inches wide!

Photographing the eclipse with a lens shorter than 300mm will leave you with little to work with. Using a 400mm lens and printing a 24-inch print will result in a 2.5-inch-wide sun. For as massive as the sun is, it is a challenge to take a large photograph of the sun. The sun will appear to move fast with a 500mm lens, too. Plan to adjust.

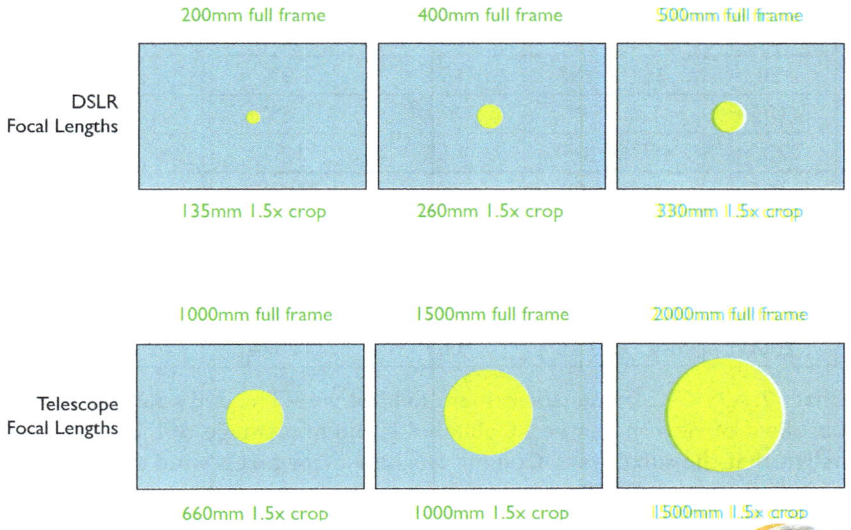

FOCAL LENGTH	FOV FULL FRAME	FF VERT. ANGLE	% OF FF	SUN PIXEL SIZE
14	104° X 81°	81°	0.7%	32.1
20	84° X 62°	62°	0.9%	41.9
28	65° X 46°	46°	1.2%	56.5
35	54° X 38°	38°	1.4%	68.5
50	40° X 27°	27°	2.0%	96.4
105	19° X 13°	13°	4.1%	200.2
200	10° X 7°	7°	7.6%	371.9
400	5° X 3.4°	3.4°	15.6%	765.6
500	4° X 2.7°	2.7°	19.6%	964.2
1000	2° X 1.3°	1.3°	40.8%	2002.5
1500	1.4° X 0.9°	0.9°	58.9%	2892.6
2000	1° X 0.68°	0.68°	77.9%	3828.4

Chart 1: Full-frame camera field of view. The 3rd column is the vertical field of view in degrees. Column 4 is the percentage of the total sensor height that the sun covers. Column 5 is how many pixels wide the sun will be on a 36MP Nikon D810. (Values are estimates)

FOCAL LENGTH	FOV CROP	CROP VERT DEG	% OF CROP	SUN PIXEL SIZE
14	80° X 58°	58°	0.9%	33.9
20	61° X 43°	43°	1.2%	45.8
28	45° X 31°	31°	1.7%	63.5
35	37° X 25°	25°	2.1%	78.7
50	26° X 18°	18°	2.9%	109.3
105	13° X 8°	8°	6.6%	245.9
200	6.7° X 4.5°	4.5°	11.8%	437.2
400	3.4° X 2°	2°	26.5%	983.7
500	2.7° X 1.8	1.8°	29.4%	1093.0
1000	1.3° X 0.9°	0.9°	58.9%	2186.0
1500	0.9° X 0.6°	0.6°	88.3%	3278.9
2000	0.6° X 0.45°	0.5°	117.8%	4371.9

Chart 2: APS-C Crop sensor camera field of view. The 3rd column is the vertical field of view in degrees. Column 4 is the percentage of the total sensor height that the sun covers. Column 5 is how many pixels wide the sun will be on a 12mp Nikon D300s. (Values are estimates)

The big challenge is the cost of the lens. Lenses longer than 300mm are expensive. They also require heavier tripods and specialized tripod heads. The 70-300mm lenses from Nikon, Canon, Tamron, and others are relatively affordable options. It is worth spending time at a local camera shop to try different lenses. Long focal-length lenses are a significant investment, especially for a single event.

To achieve a large eclipse image, you will need a long focal-length lens, ideally at least 400mm. A standard 70-300mm lens set to 300mm will show a small sun. At 500mm, the sun image becomes larger and covers more of the sensor area. The corona will take up a significant portion of the frame. By 1000mm, the corona will exceed the capture area on a full-frame sensor. See the picture in this chapter for sun size simulations for different focal lengths.

Suggested Exposures

To photograph the partial eclipse, the camera must have a solar filter attached. If not, the intense light from the sun may damage (fry) the inside of your camera. This has happened to the author. The exposure depends on the density (darkness) of the solar filter used.

As a starting point, set the camera to ISO 100, f/8, and with the solar filter on, try an exposure of 1/2000. Make adjustments based on the filter used, histogram, and highlight warning.

Turn on the highlight warning in your camera. This feature is commonly called "blinkies." This warning will help you detect if the image is overexposed or not.

Once the Baily's Beads, prominences, and corona become visible, there will only be a few minutes to take bracketed shots. It will take at least eleven shots to capture the various areas of the sun's corona and stars. The brightness varies considerably. No commercially available camera can capture the incredible dynamic range of the different portions of the delicate corona. This requires taking multiple photographs and digitally combining them afterward.

During totality, try these exposure times at ISO 100 and f/8: 1/4000, 1/2000, 1/1000, 1/250, 1/60, 1/30, 1/15, 1/4, 1/2, 1 sec, and 4 sec.

Disable camera/lens stabilization on a tripod!

Photography Time

Set the camera to full-stop adjustments. It will reduce the time spent fiddling. As an example, the author tried the above shot sequence, adjusting the shutter speed as fast as possible.

It took thirty-three seconds to shoot the above 11 shots using 1/3-stop increments. This was without adjusting composition, focus, or anything else but the shutter speed. When the camera was set to full stop increments, it only took twenty-two seconds to step through the same shutter speed sequence. Use a remote release to reduce camera shake.

Assuming the totality lasts less than two minutes, only four shot sequences could be made using 1/3-stop increments. Yet six shot sequences could be made when the camera was set to full stop steps. Zero time was spent looking at the back LCD to analyze highlights and the histogram.

Now add in the bare minimum time to check the highlight warning. It took sixty-three seconds to shoot and check each image using full stops. And that was without changing the composition to allow for sun movement, bumping the tripod, etc. Looking at the LCD ("chimping") consumed **half** of the totality time.

This test was done in the comfort of home under no pressure. In real world conditions, it may be possible to successfully shoot only one sequence. If you plan to capture the entire dynamic range of the totality, you must practice the sequence until you have it down cold. If you normally fumble with your camera, do not underestimate the difficulty, frustration, and stress of total eclipse photography.

Most importantly, trying to shoot this sequence allowed for zero time to simply look at the totality to enjoy the spectacle.

Avoid Last Minute Purchases

You should purchase whatever you think you'll need to photograph the eclipse early. This event will be nothing short of massive. Remember the hot toy of the year? Multiply that frenzy by a thousand. Everyone will want to try to capture their own photo.

Do not wait until the last few weeks before the eclipse to purchase cameras, lenses, filters, tripods, viewing glasses, and associated material. Consider that the totality of the eclipse will streak across

America. Everyone who wants to photograph the eclipse will order at the same time. If you wait until too late to buy what you need, it's conceivable that solar filters to create a total eclipse photo will be sold out in the United States. All filters sold out during the 2017 total eclipse. Whether this happens or not, do not wait to make your purchases. It may be too late.

Practice

You will need to practice with your equipment. Things may go wrong that you don't anticipate. If you've never photographed a partial or total eclipse, taking quality shots is more difficult than you think. Practice shooting the sequence with a midday sun. This will tell you if you have your exposures and timing correct. Figure out what you need well in advance.

Practice photographing the full moon and stars at night. Capture the moon in full daylight to learn how your camera reacts. Astrophotography is challenging and requires practice.

The August 21, 2017, eclipse as seen in Jackson, WY, shot with a Nikon D800 with an 80-400mm lens set to 340mm. The sun is 644 pixels wide on the 7360x4912 image.

This image is shown straight out of the camera without modification. Even with a high-quality camera and lens, photographing an eclipse is challenging.

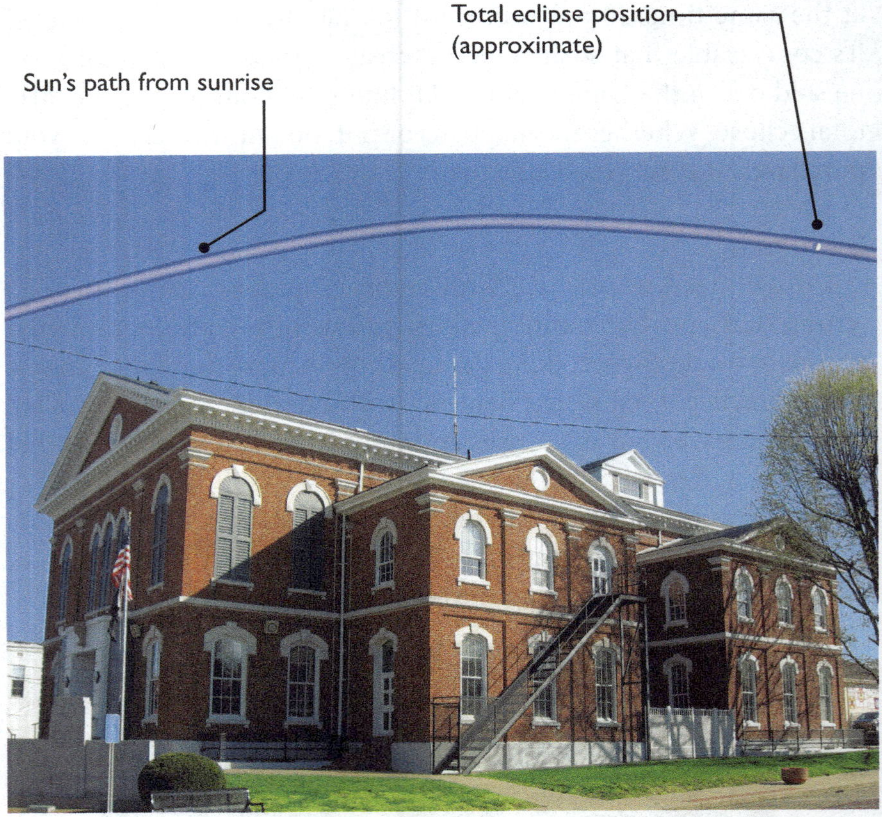

The eclipse will follow this approximate path on the afternoon of April 8, 2024.

Image of Union County Courthouse, Morganfield, Kentucky.

Note that this image is a simulation and approximation the sun's path and where the total eclipse may appear from one perspective. Refer to the eclipse position data for a more accurate location.

Viewing Locations Around Kentucky

Tens of thousands of people will travel to and around Kentucky to view the total eclipse. There are few obstructions and there is a vast amount of space to view the total eclipse from.

If the weather is questionable, seek out a new location as soon as possible. If you wait until the hour before the eclipse, you may find yourself stuck in traffic, as others will be looking for a viewing location. Be safe on the roadways, as drivers may be distracted.

This section contains popular, alternative, and little-known locations to watch the eclipse. As long as there are no clouds or smoke from fires, the partial eclipse will be viewable from anywhere in the state.

Suggested Total Eclipse View Points

Towns and Cities

- Bandana
- Boxville
- Corydon
- Henderson
- Lovelaceville
- Marion
- Morganfield
- Paducah
- Reed
- Robards
- Salem
- Sturgis
- Wickliffe

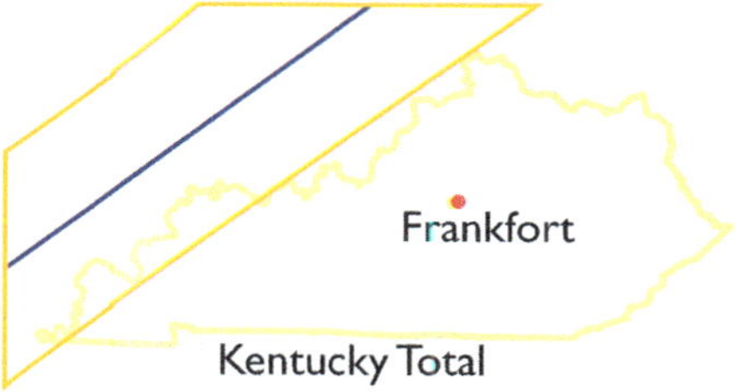

Kentucky Total Eclipse Path

Unique Locations

- Atkinson Park
- Columbus-Belmont State Park
- Diamond Island
- Green River State Forest
- Higginson-Henry Wildlife Management Area
- Ohio River
- Mississippi River
- West Kentucky State Wildlife Management Area

Bandana

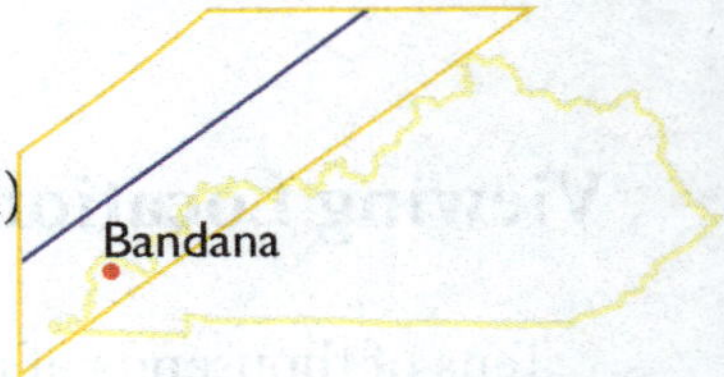

Elevation:	358 ft. (109 m)
Population:	203
Main road/hwy:	KY 358

Overview

Bandana was on the path of the 2017 total solar eclipse. There were a few locals who volunteered to host the folks seeking to watch the eclipse. Don't wait too long to contact people for hosting for the 2024 eclipse. It's a small community located twenty-two miles west of Paducah. There is an old post office operational since 1880 in the vicinity; it could be an attraction to visit if you like to go back in history. Bandana has few accommodations, but you could try a hotel in Paducah as the nearest option.

Getting There

Drive west on US 60 from Paducah for twenty miles, then turn north on Bandana Road and drive five miles to reach Bandana.

Totality Duration

3 minutes 6 seconds

Notes

Visit the Kentucky government tourism website for more information at https://kentucky.gov/Tourism/Pages/default.aspx.

Event	Time (CDT)	Altitude	Azimuth
Sunrise	6:30:00AM	0°	80°
Eclipse Start	12:42:30PM	60°	172°
Totality Start	1:59:31PM	57°	209°
Totality End	2:02:38PM	57°	210°
Eclipse End	3:18:07PM	46°	236°
Sunset	7:25:00PM	0°	280°

Boxville

Elevation:	436 ft. (133m)
Population:	3,466
Main road/hwy:	KY 141

Overview

The nearest airport to Boxville is Evansville Regional Airport, connecting major cities in the United States. The airport is around thirty-three miles (52 km) northeast of Boxville. Boxville is also well connected with road travelers. You can find many interesting things to do while you are there, like golfing (there are many golfing clubs), camping, and Camp Breckinridge Museum & Art. You can also visit wine tasting sites near Bandana. If you are with children, you can pay a visit to the Children's Museum of Evansville that would be a treat for the kids.

Getting There

Drive southeast on US 62 from Paducah, then turn north on US 60 and drive fifty-seven miles to KY 141, then continue east for eleven miles to reach Boxville.

Totality Duration

2 minutes 12 seconds

Notes

Point your web browser to https://www.kentuckytourism.com/ to find out more information about Boxville.

Event	Time (CDT)	Altitude	Azimuth
Sunrise	6:25:00AM	0°	80°
Eclipse Start	12:44:58PM	60°	176°
Totality Start	2:02:14PM	56°	212°
Totality End	2:04:27PM	56°	213°
Eclipse End	3:19:55PM	45°	237°
Sunset	7:21:00PM	0°	280°

Corydon

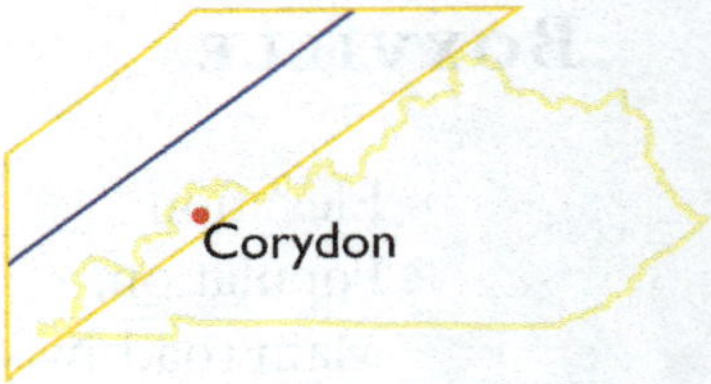

Elevation:	456 ft. (139m)
Population:	706
Main road/hwy:	US 60

Overview

Corydon is a small city and a former coal town in Henderson County, Kentucky. Corydon is connected by US Route 60 and it is also Corydon's main street. Most of the local businesses are located on the main street and this makes it easier to hunt down a place to stay or dine. A tour to Indiana Caverns will be a treat. It's a perfectly calm and serene place. Harrison County Discovery Center offers a good museum experience in Corydon. If you happen to love the art of glassmaking, you can visit Zimmerman Art Glass where you can relax and watch glassmakers at their art.

Getting There

Drive south for twenty miles from Evansville, Indiana on US 41 and continue on US 60 to reach Corydon.

Totality Duration

2 minutes 23 seconds

Notes

The Henderson County website at https://www.hendersonky.org/ will be a good starting point for your visit to Corydon.

Event	Time (CDT)	Altitude	Azimuth
Sunrise	6:24:00AM	0°	80°
Eclipse Start	12:45:19PM	60°	176°
Totality Start	2:02:26PM	56°	212°
Totality End	2:04:50PM	56°	213°
Eclipse End	3:20:08PM	45°	237°
Sunset	7:20:00PM	0°	280°

Henderson

Elevation:	407 ft. (124m)
Population:	28,657
Main road/hwy:	US 41

Overview

Henderson is one of the major cities among our shortlisted places in Kentucky. It happens to be one of the most populated Kentucky cities to experience the 2024 total eclipse. It is surrounded by small counties that also lie on the direct path of the solar eclipse. During your stay, you can visit the John James Audubon Museum and Farmer and Frenchman Winery. If you can manage to book a place in Henderson in advance, you can always visit many exciting places nearby and rest up after a long day of exploring.

Getting There

Drive south for eleven miles from Evansville, Indiana on US 41 to reach Henderson.

Totality Duration

2 minutes 31 seconds

Notes

The Henderson city website is a good starting point to find out about total eclipse activities. https://www.cityofhendersonky.org/

Event	Time (CDT)	Altitude	Azimuth
Sunrise	6:24:00AM	0°	80°
Eclipse Start	12:45:37PM	60°	177°
Totality Start	2:02:39PM	56°	212°
Totality End	2:05:11PM	55°	213°
Eclipse End	3:20:21PM	45°	237°
Sunset	7:20:00PM	0°	280°

Lovelaceville

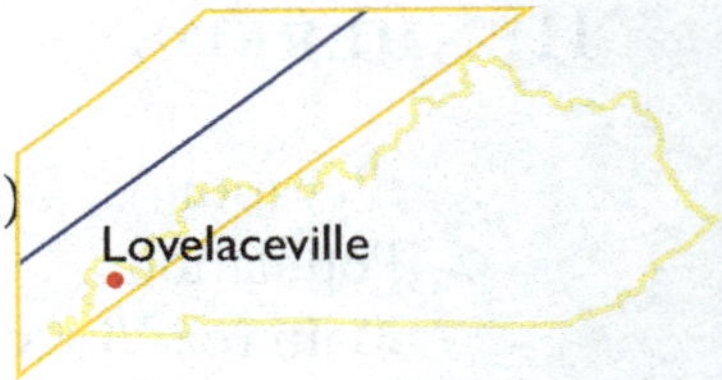

Elevation:	374 ft. (114 m)
Population:	972
Main road/hwy:	US 62

Overview

You can reach Lovelaceville either by road or air. The nearest airport is Dress Regional Airport, approximately a hundred miles from Lovelaceville. Places like Riverview mansion offer a spectacular view of the river. There are other hotels along the riverfront to choose from. Lovelaceville offers a quiet environment and serene long walks along the river. With riverfront walks and hotels offering an amazing view of the river, it could be a perfect place if you love to spend some quiet time with nature.

Getting There

Drive south on US 62 from Paducah for sixteen miles to reach Lovelaceville.

Totality Duration

1 minutes 49 seconds

Notes

Ballard County's website is a good starting point to find out more about Lovelaceville. Their site address is https://www.discoverballard-county.com/.

Event	Time (CDT)	Altitude	Azimuth
Sunrise	6:30:00AM	0°	80°
Eclipse Start	12:42:25PM	60°	173°
Totality Start	2:00:09PM	57°	210°
Totality End	2:01:58PM	57°	210°
Eclipse End	3:18:09PM	46°	236°
Sunset	7:24:00PM	0°	280°

Marion

Elevation:	594 ft. (181m)
Population:	2,930
Main road/hwy:	US 60

Overview

Marion was founded in 1842 and named after Francis Marion, a brigadier general from South Carolina. Marion is home to many internationally known museums like Ben E. Clement Mineral Museum and Crittenden County Historical Museum. Marion has the second-largest population of whitetail deer and wild turkeys. Hunters can enjoy hunting on leased private lands. During April, Marion hosts the famous Backroads Tour and Festival.

Getting There

Drive northeast from Paducah on US 62/60 for forty miles to reach Marion.

Totality Duration

34 seconds

Notes

The Marion city website will have contact information for visitors. Point your web browser to http://www.marionkentucky.org/ for more information.

Event	Time (CDT)	Altitude	Azimuth
Sunrise	6:26:00AM	0°	80°
Eclipse Start	12:44:07PM	60°	175°
Totality Start	2:02:20PM	56°	211°
Totality End	2:02:54PM	56°	212°
Eclipse End	3:19:23PM	45°	237°
Sunset	7:21:00PM	0°	280°

Morganfield

Elevation: 437 ft. (137m)
Population: 3,466
Main road/hwy: US 60

Overview

The town has a history of being the only city in the bluegrass states where Abraham Lincoln gave a political speech. Breckinridge, an infantry division camp and prisoner-of-war camp belonging to the era of World War II, is located near Morganfield. At the height of the war, the camp accommodated around 43,000 soldiers and officers. The camp also held nearly 3,000 German prisoners of war during World War II. Morganfield is surrounded by many tourist attractions like Higginson-Henry Wildlife Management Area, Mauzy Lake, and Moffitt Lake Recreation Area. During your stay, check out Camp Breckenridge Museum & Art Center or Breckinridge Golf & Country Club.

Getting There

Drive west on US 62/60 from Paducah for seventy-five miles to reach Morganfield.

Totality Duration

2 minutes 44 seconds

Notes

Visit Morganfield's website for updated total eclipse information at https://morganfield.ky.gov/.

Event	Time (CDT)	Altitude	Azimuth
Sunrise	6:25:00AM	0°	80°
Eclipse Start	12:44:54PM	60°	176°
Totality Start	2:01:53PM	56°	211°
Totality End	2:04:37PM	56°	213°
Eclipse End	3:19:50PM	45°	237°
Sunset	7:21:00PM	0°	280°

Paducah

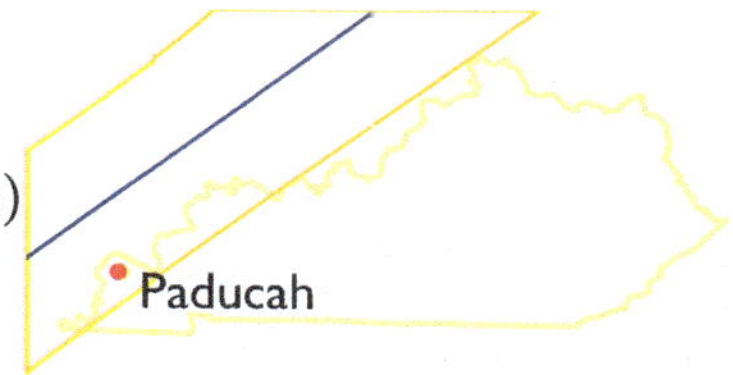

Elevation:	341 ft. (104 m)
Population:	24,944
Main road/hwy:	US 60

Overview

Paducah is the second most populated city on our list and has plenty of attractions to enjoy. In 2013 Paducah was designated as a creative city of crafts and folk arts by UNESCO. Paducah is famous for its history of steamboats, railroads, and its artist attractions like the National Quilt Museum, Paducah "Wall to Wall" floodwall murals, and the Paducah School of Art and Design. Interstate 24 runs through Paducah, and Barkley Regional Airport serves as the nearest airport to the city.

Getting There

Drive north on I-40/I-65 for one hundred thirty-seven miles from Nashville, Tennessee to reach Paducah.

Totality Duration

1 minutes 34 seconds

Notes

The Paducah city website at http://paducahky.gov/ is a great starting point to find out more about total eclipse events.

Event	Time (CDT)	Altitude	Azimuth
Sunrise	6:29:00AM	0°	80°
Eclipse Start	12:42:57PM	60°	173°
Totality Start	2:00:45PM	57°	210°
Totality End	2:02:20PM	57°	211°
Eclipse End	3:18:32PM	46°	236°
Sunset	7:23:00PM	0°	280°

Reed

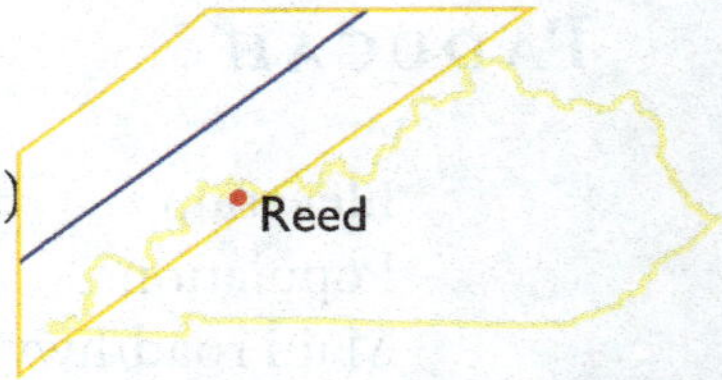

Elevation:	377 ft. (115 m)
Population:	786
Main road/hwy:	US 60

Overview

Reed has many local places that can provide a memorable experience during your stay. Take in a concert at Mojo's Boneyard Sports Bar & Grille, a famous local spot. If you want to fly in, the nearest airport is Daviess County Airport. You can visit Moore Music, a huge musical instrument store. Take some time out to visit the international bazaar offering fantastic Indian food for dining in or authentic Indian spices to take home.

Getting There

Drive south from Evansville, Indiana on US 41 and then turn east on US 60 and continue for a total of twenty-one miles to reach Reed.

Totality Duration

1 minutes 43 seconds

Notes

Visit the Henderson county website at http://hendersonky.us/ for more information on events in and around Reed.

Event	Time (CDT)	Altitude	Azimuth
Sunrise	6:23:00AM	0°	80°
Eclipse Start	12:46:01PM	60°	178°
Totality Start	2:03:26PM	55°	213°
Totality End	2:05:09PM	55°	213°
Eclipse End	3:20:40PM	44°	237°
Sunset	7:19:00PM	0°	280°

Robards

Elevation:	430 ft. (131m)
Population:	511
Main road/hwy:	KY 416

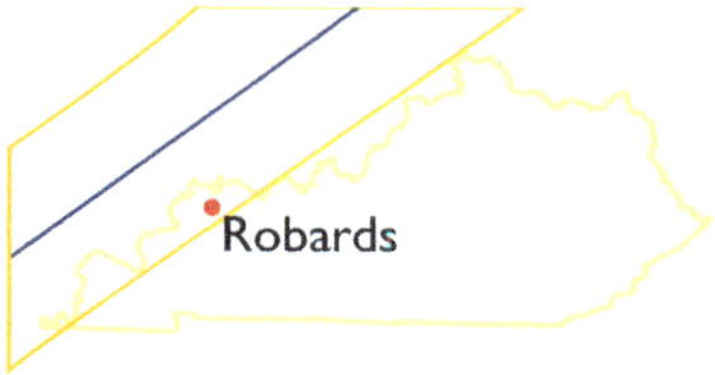

Overview

If you want to witness the 2024 celestial event in Robards, you can stay at the Reine cabin at the Farmer and Frenchman winery. It is a local attraction where you can enjoy a session of wine tasting. You can find accommodations in hotels like Super 8 by Wyndham Owensboro or Economy Inn and Suites. They offer online bookings as well, so if you decide to visit Robards, book in advance. The nearest airport is Daviess County Airport.

Getting There

Drive south from Evansville, Indiana on US 41 and continue on I-69 until you reach KY 416, then turn west until you reach Robards.

Totality Duration

1 minutes 1 second

Notes

The Kentucky tourism website at https://www.kentuckytourism.com/ is a good starting point for information on Robards and surrounding communities.

Event	Time (CDT)	Altitude	Azimuth
Sunrise	6:24:00AM	0°	80°
Eclipse Start	12:45:27PM	60°	177°
Totality Start	2:03:19PM	56°	212°
Totality End	2:04:19PM	55°	213°
Eclipse End	3:20:18PM	45°	237°
Sunset	7:19:00PM	0°	280°

Salem

Elevation:	443 ft (135 m)
Population:	752
Main road/hwy:	US 60

Overview

Salem is a part of the Paducah metropolitan area. Salem was believed to have been settled in 1800 by immigrants from North Carolina. Although Salem is a small city, there are many exciting places to visit. You can pay a visit to Clement Mineral Museum and the American Fluorite Museum.

Getting There

Drive northeast from Paducah on US 62/60 for thirty-four miles to reach Salem.

Totality Duration

1 minutes 8 seconds

Notes

Check with the Clement Mineral Museum for total eclipse activities. They can be contacted via their website at https://www.clementmineralmuseum.org/.

Event	Time (CDT)	Altitude	Azimuth
Sunrise	6:27:00AM	0°	80°
Eclipse Start	12:43:46PM	60°	174°
Totality Start	2:01:43PM	56°	211°
Totality End	2:02:52PM	56°	212°
Eclipse End	3:19:07PM	46°	237°
Sunset	7:22:00PM	0°	280°

Sturgis

Elevation:	371 ft. (113 m)
Population:	1,332
Main road/hwy:	US 60

Overview

Sturgis is a small town founded in 1886. If you are in the vicinity, visit Cave-in-Rock State Park or Camp Breckenridge Museum & Art Center. Both places attract a lot of tourists and backpackers. Venture River Water Park has water sliding and a swimming pool. This may be chilly in April, but the kids will enjoy it. If you are into wine tasting, take a tour of White Buck Winery. There are few motels by the roadside and you can also book cottages online.

Getting There

Drive northeast from Paducah on US 62/60 for sixty-four miles to reach Sturgis.

Totality Duration

2 minutes 13 seconds

Notes

Check out the floating dock near Sturgis at https://www.unioncountyky.org/caseyville-boat-dock as an option for your total eclipse experience.

Event	Time (CDT)	Altitude	Azimuth
Sunrise	6:26:00AM	0°	80°
Eclipse Start	12:44:35PM	60°	175°
Totality Start	2:01:53PM	56°	211°
Totality End	2:04:07PM	56°	212°
Eclipse End	3:19:39PM	45°	237°
Sunset	7:21:00PM	0°	280°

Wickliffe

Elevation:	420 ft. (128m)
Population:	672
Main road/hwy:	US 51

Overview

The city stands on a twelfth-century historic site of a Mississippi cultural village. There is a museum and a state-owned research center in Wickliffe to visit. While you are there, take time to visit Fort Jefferson Memorial Cross and the Wickliffe Mound State Historic Site. Both places offer history about the town and nation. Fort Jefferson was established during the Revolutionary War. The fort was built to protect the United States from the British Army. The fort also served as a Union Army Camp during the Civil War. You can find plenty of historical material in the fort's museum.

Getting There

Drive west on US 45/62 from Paducah for eleven miles, then continue on KY 286 for seventeen miles to reach Wickliffe.

Totality Duration

2 minutes 45 seconds

Notes

Wickliffe's Mound State Historic Website will have contact information for eclipse viewing at https://www.kentuckytourism.com/wickliffe-mounds-state-historic-site/.

Event	Time (CDT)	Altitude	Azimuth
Sunrise	6:31:00AM	0°	80°
Eclipse Start	12:42:00PM	57°	209°
Totality Start	1:59:16PM	57°	209°
Totality End	2:02:01PM	57°	210°
Eclipse End	3:17:47PM	46°	236°
Sunset	7:25:00PM	0°	280°

Atkinson Park

Elevation:	407 ft. (124m)
Main road/hwy:	US 41

Overview

Also in Henderson is Atkinson Park and Pool. The beautiful riverfront flows along a serene three miles of paved trail within range of the city, making it a perfectly safe place to spend outdoor family time during the event. The park has a swim center with a pool, mushroom showers, and a water slide. The admission to the pool is based on age and quite affordable. They also have a kids playground, which makes it the perfect hangout place during the event.

Getting There

Atkinson Park is located north the US 60 and US 41 interchange.

Totality Duration

2 minutes 31 seconds

Notes

Visit the Henderson website at https://www.cityofhendersonky.org/ for more event information on Atkinson Park.

Event	Time (CDT)	Altitude	Azimuth
Sunrise	6:24:00AM	0°	80°
Eclipse Start	12:45:37PM	60°	177°
Totality Start	2:02:39PM	56°	212°
Totality End	2:05:11PM	55°	213°
Eclipse End	3:20:21PM	45°	237°
Sunset	7:20:00PM	0°	280°

Columbus-Belmont State Park

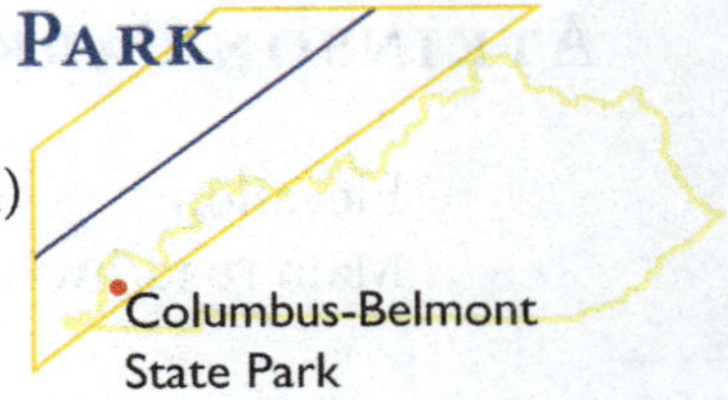

Elevation:	436 ft. (133m)
Main road/hwy:	KY 358

Overview

Columbus-Belmont State Park hosted the event in 2017 by welcoming eclipse enthusiasts and organizing solar eclipse meals for the kids. The park has the remnants of a giant chain called "Pillow's Folly" that was stretched across the river and a giant anchor weighing between four and six tons. The park also hosts the remains of a giant cannon named Lady Polk after General Leonidas Polk's wife.

Getting There

Drive south from Wickliffe on US 51 for fifteen miles, then turn west on KY 80 for five miles, and then turn south on KY 123 and continue to the park.

Totality Duration

1 minutes 36 seconds

Notes

The park's website will have updated eclipse event information at https://parks.ky.gov/parks/recreationparks/columbus-belmont/.

Event	Time (CDT)	Altitude	Azimuth
Sunrise	6:31:00AM	0°	80°
Eclipse Start	12:41:40PM	60°	171°
Totality Start	1:59:35PM	57°	209°
Totality End	2:01:11PM	57°	210°
Eclipse End	3:17:37PM	47°	236°
Sunset	7:25:00PM	0°	280°

Diamond Island

Elevation:	357 ft. (109m)
Main road/hwy:	KY 141

Overview

Diamond Island, about ten miles west of Henderson, Kentucky, offers a breathtaking view of the Ohio River. With the river flowing on both sides, it could be a hotspot to visit during the event. It has a history of massacres and pirates looting and was infamously known as a hideout for pirates during the late eighteenth century. But all that is history now. You can visit the island safely and return. The island was auctioned to the private owners in 2011.

Getting There

Drive southwest from Henderson on US 60, then continue northwest on KY 136 and KY 268 for twelve miles to reach Diamond Island.

Totality Duration

3 minutes 7 seconds

Notes

Learn more about the history of Diamond Island at https://www.landreport.com/2011/12/sold-kentucky%E2%80%99s-diamond-island/.

Event	Time (CDT)	Altitude	Azimuth
Sunrise	6:24:00AM	0°	80°
Eclipse Start	12:45:25PM	60°	176°
Totality Start	2:02:08PM	55°	211°
Totality End	2:05:16PM	55°	213°
Eclipse End	3:20:09PM	45°	237°
Sunset	7:21:00PM	0°	280°

Green River State Forest

Elevation:	374 ft. (114 m)
Main road/hwy:	US 60

Overview

Green River State Forest offers a lot of adrenaline-rush activities such as trails for hiking, biking, and horseback riding. The park also has 157 campground sites with plenty of boats to rent. The park also offers many family-friendly activities and has restaurants that offer authentic Kentucky hospitality.

Getting There

Green River State Forest is located eight miles northeast of Henderson, north of US 60 on Tscharner Road.

Totality Duration

2 minutes 28 seconds

Notes

Find more information about the forest at https://eec.ky.gov/Natural-Resources/Forestry/ky-state-forests/Pages/Green-River-State-Forest.aspx.

Event	Time (CDT)	Altitude	Azimuth
Sunrise	6:23:00AM	0°	80°
Eclipse Start	12:45:51PM	59°	177°
Totality Start	2:02:53PM	55°	212°
Totality End	2:05:22PM	55°	213°
Eclipse End	3:20:31PM	44°	237°
Sunset	7:19:00PM	0°	280°

Higginson-Henry Wildlife Management Area

Elevation:	524 ft. (160m)
Main road/hwy:	KY 56

Overview

For accommodations near Higginson-Henry Wildlife Management Area, consider Courtyard by Marriott Evansville East, Hilton Garden Inn Evansville, or maybe the Residence Inn by Marriott Evansville. The area is spread across 5,000 acres with special hunting regulations, thirty miles of a hiking-only trail, and nine miles of designated horseback riding. It used to be an army camp during World War II. After the war, the federal government sold off the agricultural land to private buyers, and the hilly part was acquired by the state to turn it into the wildlife management area.

Getting There

Drive north from Sturgis for ten miles on US 60, then turn east on US 60 Bypass and drive for one mile, then turn southeast on KY 56 to reach the Higginson-Henry WMA area.

Totality Duration

2 minutes 23 seconds

Notes

https://app.fw.ky.gov/Public_Lands_Search/detail.aspx?Kdfwr_id=132

Event	Time (CDT)	Altitude	Azimuth
Sunrise	6:25:00AM	0°	80°
Eclipse Start	12:44:53PM	60°	175°
Totality Start	2:02:04PM	56°	211°
Totality End	2:04:27PM	55°	213°
Eclipse End	3:19:51PM	45°	237°
Sunset	7:21:00PM	0°	280°

Ohio River

Elevation:	Multiple
Main road/hwy:	Various

Overview

The Ohio River offers a magnificent and serene view of wildlife. Visit and camp at one of the many towns along the riverfront. Louisville watches over much of the river. There are abundant hotels and restaurants in the town. Louisville is also the host of the famous Kentucky Derby annual festival, which takes place in the month of April. If you want to get away from the city, High Bridge, Kentucky, is a good option. It's a very quiet and tranquil town, home to the High Bridge, mimicked after the famous Brooklyn Bridge.

Getting There

The Ohio River is covered by the total eclipse path from Stanley to Wickliffe.

Totality Duration

Various depending on location. Note that portions of the river are not under the totality near Ledbetter.

Mississippi River

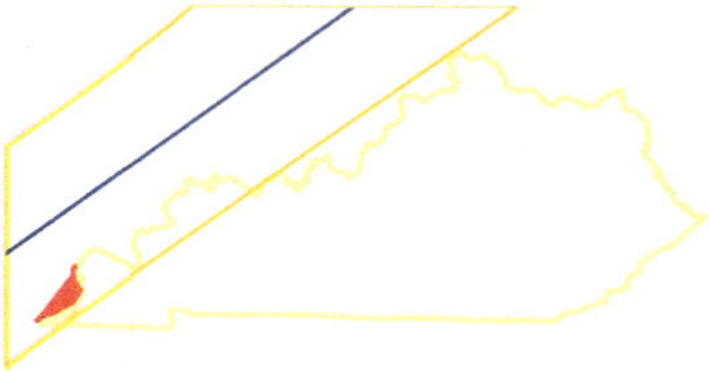

Elevation: Multiple
Main road/hwy: Various

Overview

The Mississippi River travels through Kentucky's stunning wildlife parks. You can pick a spot near Columbus-Belmont State Park or Wickliffe Mounds State Historic Site. This historic site is a twelfth-century village settlement. You can also visit the towns of Paducah and Wickliffe to find accommodations.

Getting There

The Mississippi River is covered by the total eclipse path from Wickliffe to the Kentucky, Tennessee, and Missouri border.

Totality Duration

Various depending on location. Note that portions of the river are not under the totality near Hickman.

West Kentucky State Wildlife Management Area

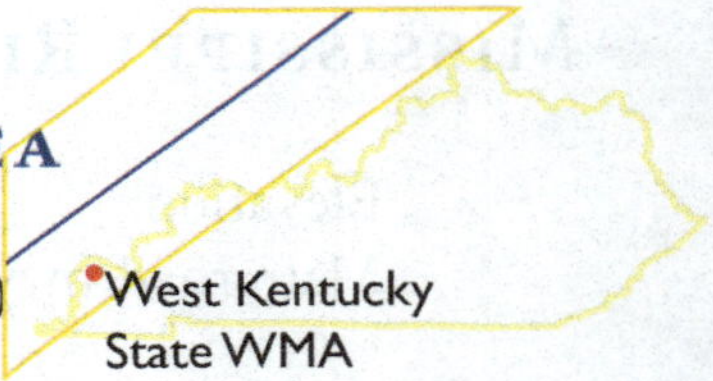

Elevation: 370 ft. (113 m)
Main road/hwy: KY 358

Overview

During November, this wildlife management area hosts the top twelve dogs from around the country in the Invitational Quail Championship Field Trial. The area supports fishing, hiking, picnicking, horseback riding, and primitive camping. There are no substantially developed areas, making this a prime location for a real nature experience for the total eclipse.

Getting There

Drive west from Paducah on KY 305 and 358 for thirteen miles to reach the West Kentucky State Wildlife Management Area.

Totality Duration

2 minutes 40 seconds

Notes

The Kentucky Department of Fish and Wildlife Resources webpage has more information about this area at https://app.fw.ky.gov/Public_Lands_Search/detail.aspx?Kdfwr_id=137.

Event	Time (CDT)	Altitude	Azimuth
Sunrise	6:29:00AM	0°	80°
Eclipse Start	12:42:39PM	60°	172°
Totality Start	1:59:54PM	57°	209°
Totality End	2:02:34PM	57°	210°
Eclipse End	3:18:15PM	46°	236°
Sunset	7:24:00PM	0°	280°

Remember the Kentucky Total Eclipse

April 8, 2024

Who was I with? ____________________

What did I see? ____________________

What did I feel? ____________________

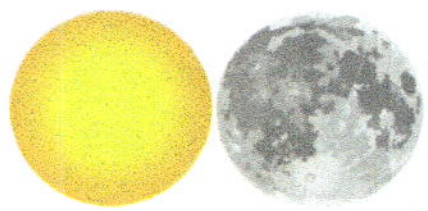

What did the people with me think?

Where did I stay?

Enjoy Other Books by Aaron Linsdau

50 Jackson Hole Photography Hotspots

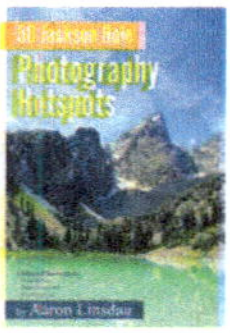

This guide reveals the best Jackson Hole photography spots. Learn what locals and insiders know to find the most impressive and iconic photography locations in the United States. This is an excellent companion guide to the *Jackson Hole Hiking Guide.*

www.sastrugipress.com/books/50-jackson-hole-photography-hotspots/

Adventure Expedition One
by Aaron Linsdau M.S. & Terry Williams, M.D.

Create, finance, enjoy, and return safely from your first expedition. Learn the techniques explorers use to achieve their goals and have a good time doing it. Acquire the skills, find the equipment, learn to camp, understand medical issues, and learn the planning necessary to pull off an expedition.

www.sastrugipress.com/books/adventure-expedition-one/

Antarctic Tears

Experience the honest story of solo polar exploration. This inspirational true book will make readers both cheer and cry. Coughing up blood and fighting skin-freezing temperatures were only a few of the perils Aaron Linsdau faced. Travel with him on a world-record expedition to the South Pole.

www.sastrugipress.com/books/antarctic-tears/

How to Keep Your Feet Warm in the Cold

Keep your feet warm in cold conditions on chilly adventures with techniques described in this book. Packed with dozens and dozens of ideas, learn how to avoid having cold feet ever again in your outdoor pursuits.

www.sastrugipress.com/books/how-to-keep-your-feet-warm-in-the-cold/

Jackson Hole Hiking Guide

Jackson Hole contains some of the most dramatic and iconic landscapes in the United States. The book shares everything you need to know to hike Jackson's classic trails with canyons, high mountains, and hidden alpine lakes. This book is an excellent companion guide to *50 Jackson Hole Photography Hotspots.*

www.sastrugipress.com/books/jackson-hole-hiking-guide/

Subscribe to Aaron's YouTube channel at www.youtube.com/@alinsdau

If you enjoyed this book, please consider leaving a five-star review and a few words on what you liked about it at your favorite online retailer.

Lost at Windy Corner

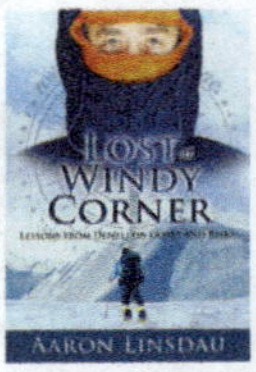

Windy Corner on Denali has claimed fingers, toes, and even lives. What would make someone brave lethal weather, crevasses, and avalanches to attempt to summit North America's highest mountain? Aaron Linsdau shares the experience of climbing Denali alone and how you can apply the lessons to your life.

www.sastrugipress.com/books/lost-windy-corner/

The Motivated Amateur's Guide to Winter Camping

Winter camping is one of the most satisfying ways to experience the wilderness. It is also the most challenging style of overnighting in the outdoors. Learn 100+ tips from a professional polar explorer on how to winter camp safely and be comfortable in the cold.

www.sastrugipress.com/books/the-motivated-amateurs-guide-to-winter-camping/

Two Friends and a Polar Bear
by Terry Williams, M.D. & Aaron Linsdau

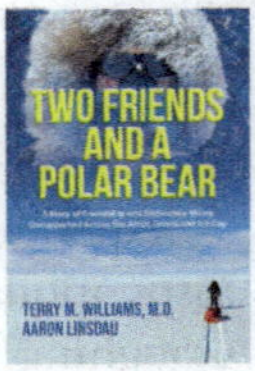

This story of friendship is about two old friends who plan to ski across the Greenland Ice Cap along the Arctic Circle in hopes of becoming one of the oldest teams to succeed.

www.sastrugipress.com/books/two-friends-and-a-polar-bear/

About the Author

Aaron Linsdau is the second American to ski alone from the coast of Antarctica to the South Pole (730 miles / 1174 km), setting a world record for surviving the longest expedition ever for that trip. He lead a 310-mile (499 km) ski expedition across the Greenland icecap along the Arctic Circle. Aaron has climbed Denali solo, crossed the Greenland tundra alone, skied across Yellowstone National Park solo, trekked through the Sahara desert, and successfully climbed Mt. Kilimanjaro and Mt. Elbrus in Russia.

Aaron Linsdau at the South Pole.

Use your smart device to scan the QR codes for website links.

Visit www.aaronlinsdau.com/subscribe to learn more about the author. Receive updates when he releases new books and shows.

Visit Sastrugi Press on the web at www.sastrugipress.com to purchase the above titles in bulk. They are available in print, e-book, or audiobook form.

Thank you for choosing Sastrugi Press.

Enjoy Other Books by Sastrugi Press

50 Florida Wildlife Hotspots by Moose Henderson Ph.D.

This is a definitive guide to finding where to photograph wildlife in Florida. Follow the guidance of a professional wildlife photographer as he takes you to some of the best places to see wildlife in the Sunshine State.

www.sastrugipress.com/books/50-florida-wildlife-hotspots/

50 Wildlife Hotspots Grand Teton National Park by Moose Henderson Ph.D.

Find out where to find animals and photograph them in Grand Teton National Park from a professional wildlife photographer. Learn techniques, timing, animal behavior, and composition to create stunning wildlife images.

www.sastrugipress.com/books/50-wildlife-hotspots/

Alaska: A Guide for the Curious by Nikki Mann & Jeff Wohl

Discover the natural world of Alaska. Find out what the plants and animals are like, how to identify them, and what the environment of Alaska is like.

www.sastrugipress.com/books/alaska-a-guide-for-the-curious/

Blood Justice by Tim W. James

Two brothers, one a preacher's son, the other an adopted would-be slave, set out in opposite directions to avenge their family's murder only to cross paths in pursuit of the killer. Book 1 of the Roger Brinkman Series.

www.sastrugipress.com/iron-spike-press/blood-justice/

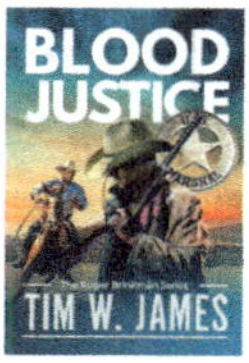

Shake Yourself Free by Bob Millsap

Learn how to overcome difficult encounters with misfortune, tragedy, and loss. Emotional recovery is a journey requiring a mindset shift. Get this book now and take control of your life.

www.sastrugipress.com/books/shake-yourself-free/

The Burqa Cave by Dean Petersen

Still haunted by Iraq, a retired soldier seeks solace teaching high school in Wyoming. He soon finds the quiet town is home to murderers, maniacs, and a boy who can see where missing murder victims are. This paranormal thriller-romance surprises readers with unexpected twists.

www.sastrugipress.com/books/the-burqa-cave/

Photography Notes

April 8, 2024

Settings

Successes

Challenges

Photography Notes

April 8, 2024

Camera

Company

Location